Single Ended Triode Amplifier
From Scratch
Richard (Dick) Whipple

Single Ended Triode Amplifier – From Scratch

Copyright © 2022 by Richard Whipple. All Rights Reserved.

All rights reserved. No part of this book may be reproduced in any form or by any electronic or mechanical means, including information storage and retrieval systems, without permission in writing from the author. The only exception is by a reviewer, who may quote short excerpts in a review.

Richard (Dick) Whipple
Visit my website at www.whippleway.com

Printed in the United States of America

First Printing: February 2022
Second Edition July 2024
Amazon Paperback

ISBN 979-8-799118-938

THE INFORMATION AND SOFTWARE CONTAINED HEREIN IS FOR EDUCATIONAL PURPOSES ONLY. IT IS PROVIDED "AS IS" WITHOUT WARRANTY OF ANY KIND, EXPRESS OR IMPLIED, INCLUDING BUT NOT LIMITED TO THE WARRANTIES OF MERCHANTABILITY, FITNESS FOR A PARTICULAR PURPOSE, TITLE, AND NON-INFRINGEMENT.

Table of Contents

Richard (Dick) Whipple 1

THE INFORMATION AND SOFTWARE CONTAINED HEREIN IS FOR EDUCATIONAL PURPOSES ONLY. IT IS PROVIDED "AS IS", WITHOUT WARRANTY OF ANY KIND, EXPRESS OR IMPLIED, INCLUDING BUT NOT LIMITED TO THE WARRANTIES OF MERCHANTABILITY, FITNESS FOR A PARTICULAR PURPOSE, TITLE AND NON-INFRINGEMENT. 2

Preface 6

Preface to the Second Edition 10

Chapter 1 - We Meet the Electron 11

Chapter 2 – The Basic Diode 17

Chapter 3 –The Basic Triode 29

Chapter 4 – The Voltage Amplifier Triode 44

    Fixed-bias 44

    Grid-Leak Bias 47

    Self-bias or Cathode-Bias 47

Chapter 5 – The Power Output Triode 53

Chapter 6 – Triode Connected Pentodes 71

Chapter 7 – Biasing the Power Output Stage 76

Chapter 8 – SET Amplifier Frequency Response 81

    Coupling Capacitor 81

    Cathode Bypass Capacitor 90

    Output Transformer 94

Chapter 9 – Example SET Amplifier Design 102

    Step 1 – Choose a power output tube. 102

    Step 2 – Choose an Output Transformer. 104

Step 3 – Choose Power Output Stage Passive Components.......................... 107

Step 4 – Calculate B+ and B++ Voltages.. 108

Step 5 – Choose a Voltage Amplifier Tube.. 108

Step 6 – Choose Voltage Amplifier Components. 109

Chapter 10 – SET Breadboard Testing 113

Chapter 11 – Power Supply Design 117

Chapter 12 – Circuit Refinements 126

Add Volume Control... 126

Add Stopper Resistors.. 128

Add Audio Bypass Capacitor to Power Supply 129

Chapter 13 – Printed Circuit Board Design 130

Electrical Safety.. 130

Printed Circuit Board Sources 131

PCB Design Steps.. 131

Routing Tips 132

External Component Connections 134

Schematic and PCB Files 134

Chapter 14 – Chassis Layout, Construction and Testing 135

Chapter 15 – Wrapping-Up 152

Appendices... 153

Appendix A - Derive Maximum Power Transfer – Linear Case 153

Appendix B – Derive Max Power Transfer – Non-Linear Case 154

Appendix C – Non-Linear Distortion 158

Harmonic Distortion.. 158

Intermodulation Distortion... 159

Appendix D – Calculate Second-Order Distortion 161
Appendix E – Frequency Distortion 165
Appendix F – Transient Response 167
Appendix G – The Miller Effect 168
Appendix H – Resistance vs. Impedance.. 169
Appendix I – Voltage Amplifier Design – General 171
Appendix J – SET Amplifier Parts List 174
Appendix K – ExpressPCB SET Amplifier Schematic...................................... 178

Preface

My first amplifier used push-pull 6BQ5 vacuum tube pentodes. By the late 1970s, I had switched to solid-state amplifiers, lured by the power and low distortion specs these amplifiers publicized. As time passed, I detected something different about the sound, but I couldn't put my finger on it. Over time, I would hear or read of others, also noticing a difference.

Occasionally, I would come across A-B listening tests where listeners preferred tube amps over solid state. Specification-wise, the solid-state amps were superior, but listeners often said tube amps sounded "better." Such test results, clouded by such subjective responses, carried little weight in the marketplace and failed to sway people away from solid-state amps.

Electric guitar enthusiasts were one of the only holdout groups. With almost religious zeal, they stayed with tube amplifiers over the years. You heard it said that there was something about the sound from tube amplifiers that complemented the guitar's acoustic properties. Again, this is a highly subjective conclusion, but it is one that kept the tube amplifier market alive.

I recall eating at a café in Southwest Texas where a country western singer was performing. I noticed he had a classic tube-type guitar amp, and during his down time, I had a look. I'll never forget seeing a row of at least ten 6L6s lined up through a slit in the back of the amplifier-speaker cabinet. In speaking to him later, he said he would never trade the sound of his tube amp for solid-state "anything"!

My Road to Damascus experience with tubes came a few years later when I acquired quite literally a garage full of old electronic gear in an auction. Among the old electronic treasures were boxes of NOS (New Old Stock) tubes, including twelve 1625 industrial power amp tubes. These were the 12-volt heater version of the extremely popular 807 pentodes used extensively in Amateur (HAM) Radio.

The idea hit me, mostly for old times' sake: why not build a tube amplifier? My first thought was to go push-pull with the 1625s. I began researching designs on the internet when I came across an article on the advantages of single-ended triode (SET) amplifiers.

Two things negatively struck me. First, SET amplifiers were generally low power, often 2 watts of less. I had lived in a world of many tens of watts and could not imagine my Bozak bookshelf speakers working at such low power levels.

Second, there was an almost paranoid fixation among SET designers against negative feedback. Wasn't negative feedback a "good thing"? Didn't it extend the frequency response and lower distortion? Of course, it did. SET promoters were expecting me to accept 5% harmonic distortion when solid-state amps with copious amounts of negative feedback were claiming distortion in the few tenths of a percent. No way, I thought!

I wrestled with these issues for a while, then finally decided I could settle the matter by measuring the power delivered to my Bozak speakers by my 40-watt solid-state amplifier. If it were 8 or 10 watts, I could let the SET idea drop. I proceeded to measure the peak-to-peak voltage at my normal listening level and discovered, to my amazement, that the power level was a mere 1-watt average!

What I had not considered was that the Bozak bookshelf speakers were amazingly efficient. They required little power to produce normal sound levels. Of course, the teenager next door might disagree with what I call "normal" listening levels, and there was the quite legitimate argument that the amp needs more power when reproducing sharp, loud sounds, like thunderclaps.

While this eased my power level concerns, I still questioned the SET amplifier's elevated level of harmonic distortion. Researching this question, I learned that triode distortion was mainly even harmonics, mostly second harmonic. I also learned that even-order harmonic distortion is far less noticeable to the average listener than odd-order distortion. In fact, sources I read suggested that people find second-order distortion as high as 5% barely discernable and rarely objectionable.

Was it possible that second-order distortion "colored" the analog signal in a way that was "pleasing" to the listener's ears? Perhaps its flattening effect could somehow explain the subjective preference for SET amplifiers. Or, just maybe, some listeners favored the sound they "liked" over any requirement for authentic reproduction?

I decided it was time to make a breadboard of the SET design and see for myself. Using a 12AV6 voltage amplifier and triode connected to a 1625 power pentode (screen grid connected directly to the plate), I threw together a working SET amplifier. I used a vintage high-fidelity output transformer I had acquired at the auction and purposely avoided negative feedback when creating the design.

When it was up and running, a 0.5 vpp input produced 2 watts output with slightly less than 4% total harmonic distortion. And the sound? Well, I would not be authoring this book if I had been disappointed with the sound!

At this point began a lengthy period of tube amplifier research culminating in breadboarding several SET amplifier configurations. I tried a variety of power triodes and pentodes, eventually settling on the 1625s that I acquired in the estate sale for my entertainment center amplifier. The 1625 is an industrial power pentode and a 12-volt heater version of the extremely popular 807 pentode used extensively in Amateur (HAM) Radio. I designed and constructed a 1625-based SET amplifier that today is performing superbly with my Bozak bookshelf speakers at the heart of my entertainment center!

For the SET amplifier construction project later in this book, I chose a more readily available output tube, the 6BQ5. I also chose to go with a printed circuit board design that would be easier for the novice DIYer to build than hand wiring.

Being a "From Scratch" book, I start with electron flow in a vacuum, develop the theory of vacuum tube diodes and triodes, explore SET amplifier design, and finish with the SET amplifier construction project using triode-connected 6BQ5s.

Subjectivity is what it is, and I cannot guarantee that you find a SET amplifier better than a solid-state amplifier. The best I can do is lay out the "From Scratch" path to building a SET amplifier, and then you can decide for yourself.

Preface to the Second Edition

Not having a proofreader apart from myself, the first addition of this book contained a considerable number of typos and grammatical issues. Although none to my knowledge interfered with the technical comprehension of the text, I thought it was best to put it right. Rather than finding a human proofreader, I signed up for the premium version of Grammarly and put it to work. It made a host of suggestions (into the hundreds), which was very disheartening. Nevertheless, I trudged along and took many of its suggestions, producing what I hope is a more readable book.

I also set up an email account should you want to communicate with me. I appreciate your purchasing my book and look forward to your comments and suggestions. Email me at wwprojects@whippleway.com.

Chapter 1 - We Meet the Electron

We begin our "From Scratch" discussion of vacuum tubes with the heart of all electrical theory, the electron. Without the free-wheeling electron and our ability to control it, vacuum tubes and all the other electronic wizardry we enjoy would not be possible. So, for our first chapter, we trace the early history of electricity and the discovery of the electron.

People had observed electrical effects and philosophized about them from early Greek times. An example was the effect produced when a person rubbed amber with fur. After rubbing, amber attracted small bits of hair and straw. By the 17th century, experimenters theorized that rubbing certain objects transferred "something" from one object to another. The "something" became known as "electrical charge."[1]

With a closer look at an electrical charge, experimenters observed that once charged, objects attracted or repelled other charged objects. This behavior led to the suggestion that there were two kinds of electrical charge, eventually designated as *positive* and *negative*. The observation became the often quoted, "Like charges attract; unlike charges repel." Still, it was unclear just what was physically moving between objects to charge them electrically.

In time, experimenters devised clever machines to mechanically separate charges for experimentation. These included the Holtz, Wilmshurst, and Pidgeon machines. Additionally, they developed devices like the Leyden jar that could store charges created by these machines. The capability to easily generate and store charge and then transfer it from one object to another in a controlled manner permitted experimenters to explore electrical effects better. They called moving charge *electrical current* because it appeared to behave like a fluid flowing between oppositely charged objects.

By the end of the 17th century, experimenters had discovered a second method of separating charge. Alessandro Volta found that placing an electrolyte (cloth soaked in brine) between zinc and copper plates caused opposite electrical

---

[1] During the 17th century, experimenters of electrical phenomena coined the word "electricity" from the Latin "electricus," meaning "like amber."

charges to accumulate on the metal plates. He called his invention a voltaic pile, what today we call a battery. With time, the voltaic pile developed into a more reliable source of charge than static electricity machines.

Perhaps the greatest advantage of the voltaic cell was that it provided a nearly continuous and steady supply of charge for experimenters studying current flow. From experiments using voltaic cells, two theories emerged. The first theory proposed that there were two types of fluids, one positive and one negative. A neutral body had equal amounts of each fluid. Upon separation of charge, a body had more of one than the other. During conduction, the flow was such that positively and negatively charged fluids mixed equally, neutralizing the charge.

Next came the single-fluid theory that reduced the number of fluids to one. The single-fluid theory proposed that only positively charged fluid flowed from one body to another. A positively charged body had an excess of positive fluid; a negatively charged body had a deficiency. Discharging was simply the action of positive fluid passing through a conducting medium to neutralize a negatively charged object.

During the 17th and 18th centuries, scientists studied current flow in various mediums, including gases at low pressures. In glass tubes containing low-pressure air, the passage of electrical current produced colorful displays of light. Probably because the displays appeared like the Aurora Borealis, the Northern Lights, scientists called them *Aurora Tubes*.

In 1879, William Crookes experimented with aurora tubes such as the one shown below.

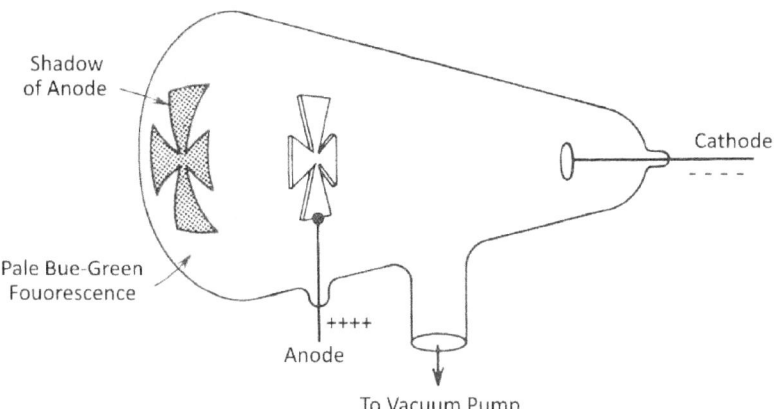

A vacuum pump reduced the air pressure inside the glass tube to an extremely low level. With the cathode negatively charged and the anode positively charged, current flowed from the cathode to the anode in a fluidic beam. As the negatively charged beam reached the anode, part of it struck the positively charged anode while the rest passed by. The result was a shadow of the anode cast in blue-green fluorescence on the end of the glass tube.

This effect was not unlike the effect of shining a light beam from the cathode toward the anode, casting a shadow on the tube end. Perhaps for this reason, Crooke referred to this fluidic beam as a *cathode ray*.

The Crookes Tube shadow hinted that the fluid consisted of particles, some of which were moving fast enough to fly by the anode and strike the glass in such a way as to cast a shadow.

To test his cathode ray theory further, Crooke devised the experiment illustrated below.

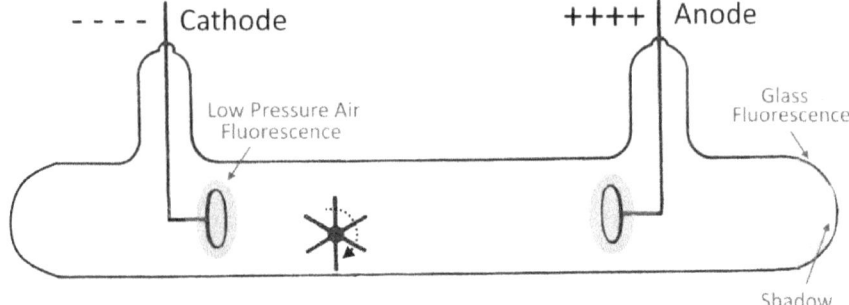

Again, the anode and cathode of a low-air-pressure aurora tube were excited with positive and negative charges. As the cathode ray passed the paddle wheel, some of the conjectured particles appeared to strike the top of it, causing it to turn. This effect, in Crookes' mind, was more evidence of the particle nature of cathode rays.

A subsequent study of the paddle wheel effect showed that the particles, electrons as it turns out, did not have sufficient momentum to turn the wheel. A different phenomenon was at work: the *radiometric effect*. Electrons striking the front side of the paddle heated it slightly above the temperature of the opposite side. The temperature difference in the low-pressure gas set up an imbalance of force that turned the wheel, albeit in the same direction as if electrons were impacting it.

We should note that Crookes' particles, whatever they were, were traveling negative to positive, opposite the direction of the prevailing 17th-century single fluid theory. Crooke correctly conjectured from this that cathode rays consisted of negatively charged, rather than positively charged, particles.

In 1880, Thomas Edison inadvertently took the single fluid idea one step further. While trying to find out why the insides of his newly developed light bulbs were darkening with time, he installed a small metal plate next to the filament.

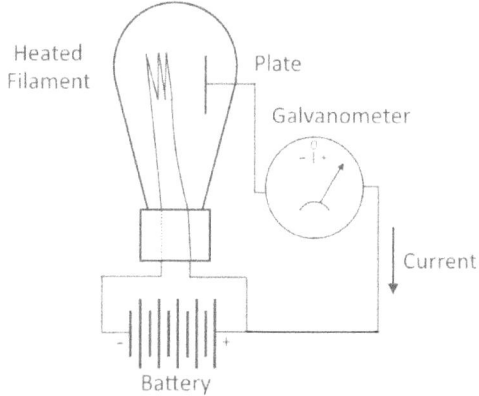

He then connected the plate to the positive side of the filament and found that a small electrical current flowed from the heated filament to the plate circuit. On the other hand, no current flowed with the plate connected to the negative side of the filament. He dismissed the result at the time as of little interest because it did not appear to relate to the glass darkening problem.

While Edison attributed little importance to his discovery, other experimenters recognized the effect as potentially important, calling it the *Edison Effect*.

Among European experimenters of the Edison Effect was W. H. Preece in England. In the mid-1880s, he measured and analyzed the volt-ampere characteristics of an Edison Effect apparatus. Little did he know that he was on the verge of discovering the diode vacuum tube!

In the early 1890s, another Englishman, J. A. Fleming, attempted to explain the mechanism of electrical conduction in low-pressure air. While he recognized that particles in the current were negatively charged, he failed to identify them as sub-atomic. He hindered his thinking by the then-prevailing view that atoms were indivisible.

The final piece of the puzzle fell into place when, in 1897, J. J. Thomson, while studying cathode rays, identified Crooke's particles as sub-atomic and negatively charged. At the time, he called them "corpuscles". We know them now as *electrons*. He determined they were over 1000 times lighter than the hydrogen atom and were the same regardless of what atom they came from. For all intents and purposes, Thomson had established the fact that current flow in evacuated glass tubes consisted of electrons!

It was Irish physicist George Johnstone Stoney who coined the term *electron*. In 1891, he was the first to suggest that there was a "single definite quantity of electricity," which he called an "electron." He combined the word *electric* with the Greek "ion" from *Ei-*, "to go."

By the start of the 20th century, the knowledge of Thomson's electron current flow in a vacuum, combined with Preece's careful analysis of the Edison Effect, heralded the beginning of the modern era of vacuum tube technology.

In the next chapter, we explore the early days of vacuum tubes and the development of supporting electrical theory.

Chapter 2 – The Basic Diode

In the previous chapter, we saw that one could explain the Edison Effect in terms of electron current flow from the negatively charged filament to a positively charged plate. Consider the basic Edison Effect device with its associated circuitry below.

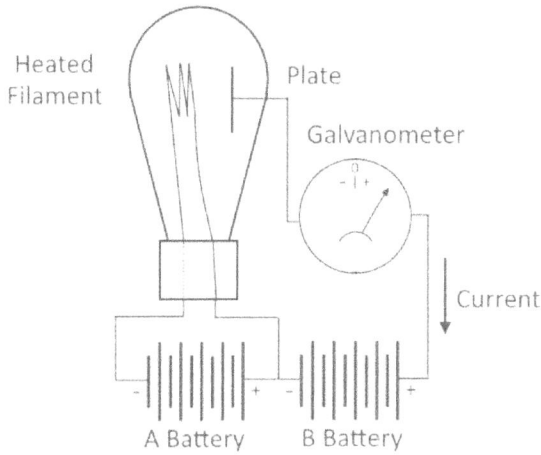

An "A" battery heats the filament while a "B" battery supplies electrons that move between the negatively charged filament and positively charged plate. A galvanometer connected in series with the plate indicates electron current flow by the movement of its needle away from the center zero position.

An important characteristic of the Edison Effect device is that current flows only when the plate is positive with respect to the filament. With the filament positively charged and the plate negatively charged, no current flows. We say that such a device exhibits a *unidirectional characteristic*.

In the early 1900s, developers of wireless telegraphy needed a unidirectional device. Here is why.

In the 1860s, James Clerk Maxwell predicted that electromagnetic waves could travel through space. In 1886, Heinrich Hertz demonstrated the existence of these *radio waves,* and by the turn of the 20th century, Guglielmo Giovanni Marconi took the next step and introduced wireless telegraphy.

A wireless radio station capable of sending and receiving Morse Code consisted of a radio transmitter and a radio receiver. An integral part of the radio receiver was a detector device that would pass electrical current in only one direction.

Sound familiar?

In common use at the time were two unidirectional devices, the *Branly coherer* and the *catwhisker crystal detector*. Both required careful adjustment and were not suitable for mobile situations like those used on ships.

British engineer and physicist J. A. Fleming was aware of the weaknesses of these detectors and saw the possibility of using an Edison Effect device as a replacement. He developed a version of the Edison Effect device that he called the *Fleming Valve*. Besides providing the needed unidirectional current flow, it did not require careful adjustment and would work equally well in all situations, both mobile and stationary.

Fleming's use of the term "valve" was in keeping with the idea of a device used to control "flow." In his case, the Fleming Valve controlled current flow much as a mechanical one-way valve controlled liquid flow. The preferred term in England, even today, is *valve*. In the U. S., proponents eventually settled on the term *vacuum tube*, referring perhaps to the tube-shaped metal or glass envelope that contained the tube's elements. Either way, Fleming's invention ushered in a whole new electronics industry on both sides of the Atlantic!

The term *diode* later came into use based on a tube having two elements (filament and plate). "Di" is Greek for "two," and "ode" means "way" or "path."

The vacuum tube diode was a precursor to the triode (three elements) and other vacuum tube devices that eventually found their way into a host of consumer, industrial, and military applications. In fact, vacuum tube technology was the driving force behind the electronics industry during the first half of the 20th Century!

Before going any further, we should take a closer look at how the diode works, which lays the foundation for understanding how vacuum tubes work.

Electron current flow in a vacuum tube diode is not as simple as it might seem at first. Let's first look at the simple case of two parallel plates in air, as shown below.

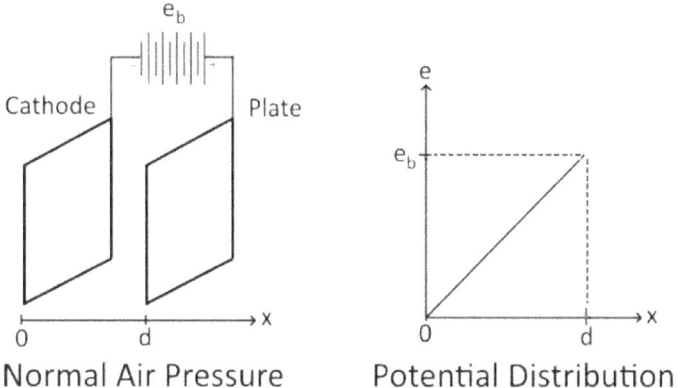

Assume a voltage e_b is impressed across them. At ambient air temperature, impacts with air molecules just off the surface impede surface electrons of the negatively charged plate from escaping. So, in this case, no free electrons exist in the space between the plates, and the electron current is zero.

Because there are no electrons in the space between the plates, the electromotive force or potential between the plates varies linearly from zero on the negative plate to e_b on the positive plate. On the right in the figure above is a potential distribution graph showing the potential as a function of distance "x" from the negative plate.

Now, suppose we place the plates in a high vacuum, as shown below.

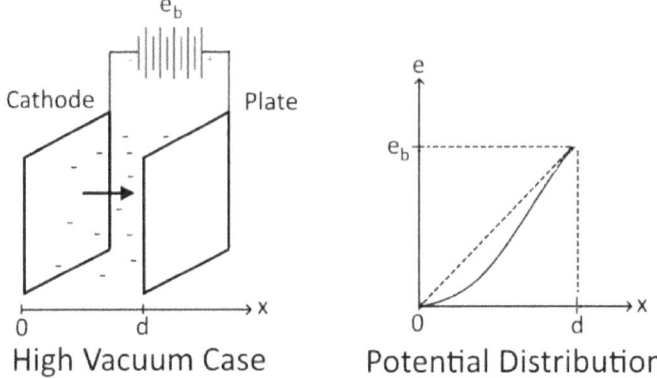

High Vacuum Case   Potential Distribution

Without interference from air molecules, electrons on the metal surface escape from the negatively charged plate. Once between the plates, an electromotive force acts on them, causing acceleration toward the positive plate. Thus, we have electron current flow between the plates.

That seems straightforward enough until we consider that the presence of negatively charged electrons in the space between plates has the effect of decreasing the potential within the space. The figure on the right above shows the *space charge* effect on the potential distribution. Note that, instead of the linear shape as before (the dotted line), the potential sags because of the presence of the space charge electrons. The resulting decreased potential between the plates reduces the electromotive force on the electrons and weakens electron current flow.

To explore the effects of space charge further, consider the vacuum tube diode circuit below.

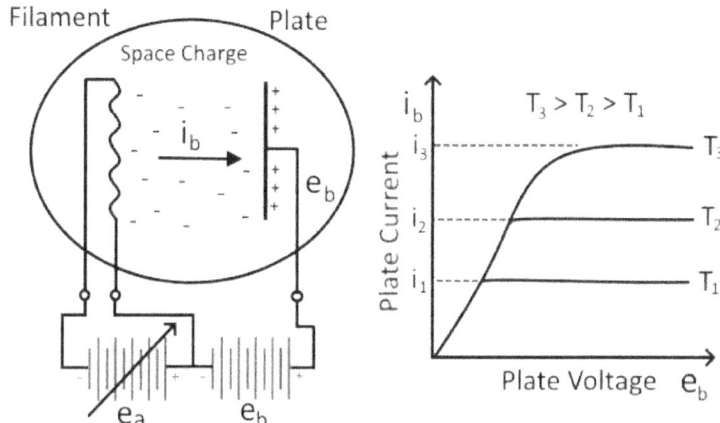

Rather than a flat plate serving as the negative source of electrons, we introduce a heated wire or *filament*. Heat increases the energy of the electrons on the surface of the filament, causing more to leave. By varying filament voltage e_a, we can control the temperature of the filament. The higher the temperature, the greater the availability of free electrons leaving the filament and moving into the space charge region. But, as the region gets more crowded with electrons, the space charge potential drops across the filament-plate region and the current i_b begins to level off at some i_s that we call the *saturation current*. For increased filament temperatures $T_1 < T_2 < T_3$, we see increased saturation currents i_1, i_2, and i_3.

We attribute the theoretical work that explained this behavior to O. W. Richardson, another in the line of British scientists studying electron flow in high vacuums. He won a Nobel Prize for his work in the emission of electricity from hot bodies. The law known as *Richardson's Equation* looks like this.

$$i_s = AT^2 e^{-\frac{b}{T}}$$

Where "i_s" is the saturation current, A and b are constants, and T is the absolute temperature of the *hot body*.

To perform correctly, we must operate the vacuum tube diode within a region below the saturation current. In the early days, engineers used vacuum tube diodes primarily for detectors in radios where voltages and currents were small. Space charge and saturation current did not come into play. In time, with the advent of AC power, vacuum tube diodes were used as rectifiers in power supplies. Rectifier applications needed high current handling capability, i.e., higher and higher saturation currents.

By choosing the right filament material, operating temperature, and physical size and layout of filament and plate, engineers raised the saturation current limit to meet the needed higher rectifier current. Below saturation current, the availability of space charge provided a kind of electron reservoir to supply high peak currents that otherwise could cause sputtering, arcing, and damage to the filament. The space charge near the filament also acted as a shield to prevent positive ions[2] from bombarding and damaging the filament.

Another factor in increasing current handling was improvement in the electron emissivity of the filament, that is, increasing the availability of free electrons at the surface of the filament. The base metal used initially for most filaments was tungsten. In a high vacuum diode, tube designers could heat tungsten to 1000-1100°K without significant deterioration over time. At such a temperature, electron emission was good but not spectacular.

In time, they found that the addition of 2% thorium to the tungsten base metal led to a huge increase in emissions. A *thoriated tungsten filament* could operate at a lower temperature while still supplying the needed current. The lower temperature extended tube life significantly. In addition, thoriated filaments were less affected by large electrical potentials, such as those in access to 500 or more volts. This characteristic permitted their use in high-current power supplies.

Another advancement came with the discovery that tungsten filaments coated with a mixture of strontium and barium oxides showed greatly improved electron emission, which was even better than thoriating. The limitation,

---

[2] There are always some positive ions present even in high vacuum tubes straight from the factory. In addition, absorbed gases from the tubes structure and glass walls are inevitably released as the tube ages.

however, was their inability to function at high voltages. Nonetheless, oxide-coated filaments became the chosen material of most receiving tubes where lower voltages (generally less than 400 volts) were commonplace.

Whatever the filament makeup, the resulting plate voltage/current characteristic of a vacuum tube diode at operating temperature looks like this.

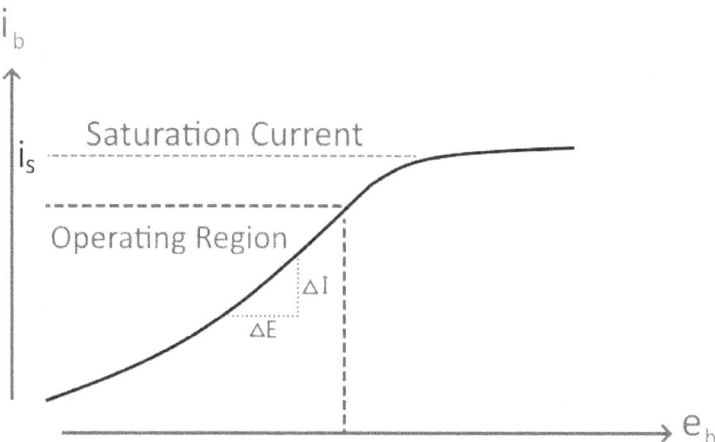

For a given filament temperature, increasing plate voltage e_b increases plate current i_b with a slightly upward trend until the space charge reserve is exhausted, at which point. The current begins to flatten out.

An ideal diode would be a short circuit when e_b is positive. A vacuum tube diode behaves more like a low-value resistor. In fact, if we calculate the ratio of a small voltage change (ΔE) to the resulting current change (ΔI) near the DC operating point of the tube diode, we can find the diode's equivalent forward resistance, R_f. For modern tube rectifier diodes, R_f hovers in the 20–75-Ω range.

If we examine i_b in the saturated region, we note that it is not flat as predicted by Richardson's equation. Instead, i_s slowly rises as e_b increases, suggesting that something is happening in the extended saturation region apart from the processes covered by the Richardson equation.

Subsequent investigation identified two processes that affect saturation current i_s in the extended saturation region. The first is the Schottky Effect, named after German physicist Walter Hans Schottky. It adds temperature dependency as well as a factor that depends on the filament material. The latter factor is significant because the Richardson Equation does not depend on the filament material.

The Schottky Effect is arrived at theoretically by considering that the electron has wave-like properties that go beyond its behavior as a particle with mass. The additional effect of voltage e_b on i_b in the extended saturation region is due to electrons exhibiting these wave-like properties.

The so-called *Field Effect* or *Fowler-Nordheim Tunneling* results in increased current with an increased electric field when operating in the extended saturation region. While the effect is slight under normal circumstances, it can produce damaging arcs at sharp points and edges within a poorly designed vacuum tube.

The point of bringing the Quantum Theory up is that we recognize that in dealing with vacuum tubes, assuming that electrons are simply particles with mass that behave like projectiles is not always adequate to explain their behavior fully. Another example of quantum theory applied to electron behavior is Einstein's equation for mass, which is related to the speed of light.

$$m_{rel} = \frac{m}{\sqrt{1-\frac{v^2}{c^2}}}$$

m_{rel} is the relativistic mass of a particle, m is the rest mass, v is the speed of the particle, and c is the speed of light about 3×10^8 meters/second or 186,000 miles/second. As a particle approaches the speed of light, its mass increases.

So far, we have ignored this effect in tubes because the speeds attained in the short distance from the filament to the plate are well below the speed of light. However, there are other vacuum tube applications where this may not be the case, and we must consider relativistic mass. Fortunately, these are beyond the scope of this book, so we trudge along, ignoring m_{rel} for now!

But I digress, back to the diode operation. We choose the *Operating Region* of a diode rectifier tube so that the maximum plate current falls below the saturation current. Also, the operating region must consider the plate's heat-radiating ability and the consequent rise in temperature. Too high a plate temperature can result in diminished performance and even tube failure[3].

Two physical processes heat the tube's plate. First, the kinetic energy of the fast-moving electrons striking the plate converts to heat energy. The power carried to the plate by the electrons is equal to plate voltage (in volts) x plate current (in amps), with the result of dissipated power (in watts).

Second, heat radiated from the filament adds to the heating of the plate. While we cannot readily control heat rise due to the filament, we can limit plate heating due to electron bombardment. By limiting the product of plate voltage and plate current, we keep plate heating to a manageable level.

In summary, these factors determine the *operation limit* of a vacuum tube rectifier diode: (1) maximum plate dissipation and (2) keeping plate current less than saturation current i_s.

In time, the separated filament and flat plate design gave way to a more commercially viable, enclosed filament design such as the one shown in the figure below.

---

[3] In power supply service, an overheated plate can source enough electrons to cause reverse-conduction when the plate voltage reverses. Not a desirable characteristic for a unidirectional device!

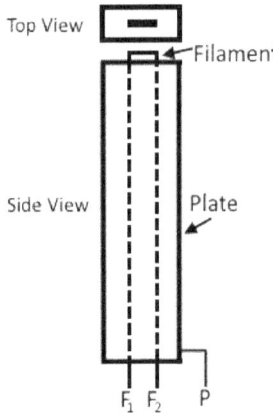

The plate consisted of a hollow metal (usually nickel) container with a rectangular cross-section. The filament passed up and down through the central opening, as shown. To handle more power, designers increased the surface area of the plate with fins that increased heat dissipation capability.

The 5U4-GB is a modern filament dual rectifier diode that utilizes this design. The 5U4-GB serves as a rectifier in high voltage (200-400 volts) and large current (100-200 ma) power supplies. Below is its plate voltage/current characteristic, schematic pin diagram, and image.

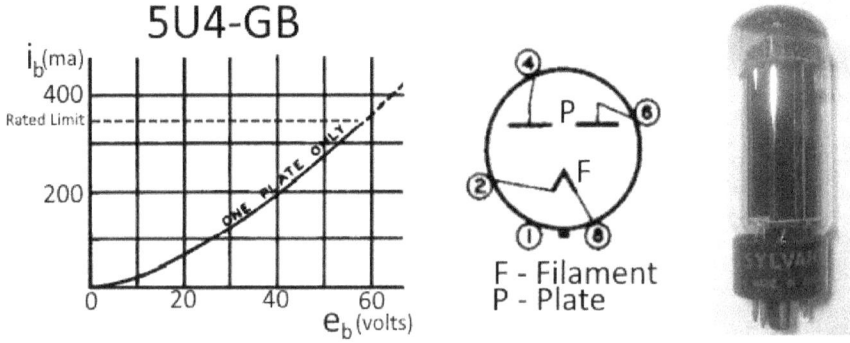

In the region below the *Rated Limit* indicated, the 5U4-GB operates well below its saturation current as well as its maximum plate dissipation. Added to the rectangular plate are fins that increase the plate's surface area and the tube's ability to radiate away heat. See the figure below.

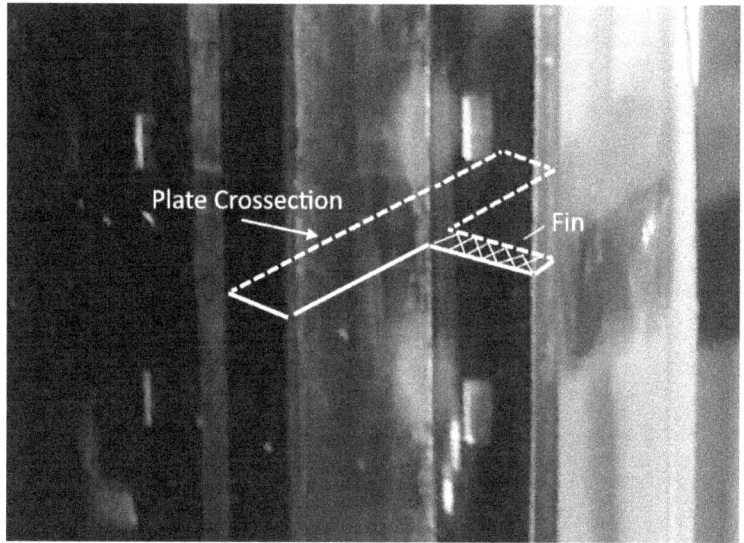

The 5U4-GB filament is oxide-coded. The large current capability (up to 350 ma) with a low voltage drop (<60 volts) made it ideal for rectifier use in vacuum tube devices such as television sets and power amplifiers.

Maintaining a vacuum tube rectifier diode in its operating range requires controlling the filament temperature[4]. So long as the filament voltage is within 10% of its specified value, we can assume that the temperature is within the desired range. The 5U4-GB, for example, has a nominal filament voltage of 5.0 volts, so 4.5 volts to 5.5 volts is theoretically acceptable.

Our journey so far has taken us from the Edison Effect in a light bulb to a modern dual-diode rectifier tube. We made each step along the way possible by the movement of electrons in a high vacuum. In the next chapter, we explore a

---

[4] The operating temperatures are as follows: tungsten 2500-2600°K (white hot); thoriated tungsten 1900-2100°K (yellow hot); and oxide-coated 950-1050°K (dull red hot).

way to further control electron flow by the addition of a *control grid* placed between the filament and plate.

Chapter 3 – The Basic Triode

In the early 1900s, American engineer Lee De Forest experimented with a Fleming diode by inserting a zig-zag wire between the filament and plate. See the sketch of De Forest's experimental vacuum tube below.

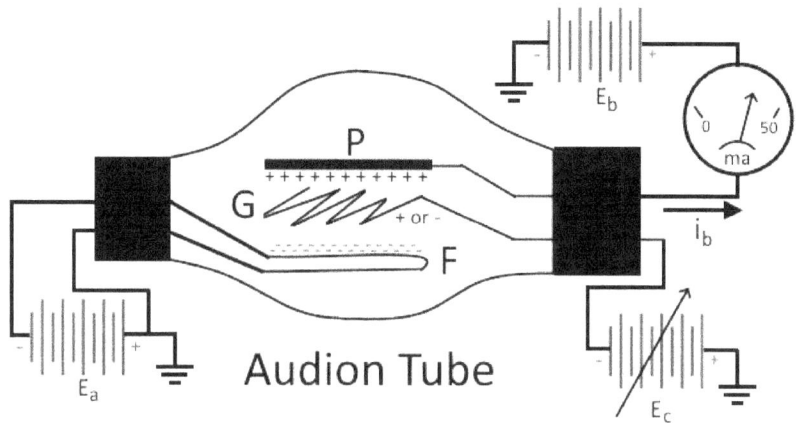
Audion Tube

De Forest called the zig-zag wire a "grid," likening it to the gridiron lines on an American football field. With the grid in place, he could control plate current by varying grid voltage E_c.

A positive voltage on the grid pulled additional electrons from the space charge region near the filament. The grid captured some of these electrons, but the majority flew by the grid and landed on the plate, increasing plate current i_b. A negative voltage on the grid repelled electrons and reduced the number, leaving the filament and space charge region for the plate. This effect, In turn, caused the plate current to decrease.

De Forrest sought to take advantage of this effect by placing a suitably sized external resistance in series with the plate circuit. The change in plate current produced a change in plate voltage <u>larger</u> than the change in grid voltage, which permitted *De Forest's Audion tube's* use as a voltage amplifier. More specifically, a time-varying electrical voltage, such as an audio signal applied to the grid, produced a duplicate, inverted, and larger voltage output at the plate!

Audion's audio amplification capability led to its first widespread use in long-distance telephone repeater circuits. As the audio signal traveled down the telephone wires, its level decreased. An Audion amplifier placed at a strategic point would bring the audio signal back to a suitable level for further transmission.

Another characteristic of the Audion was its capability to control power in the plate circuit with little power expended in the grid circuit. If the voltage between the grid and filament remained negative, only an exceedingly small grid leakage current flowed (less than a few microamperes). Thus, little grid drive power could potentially control substantial amounts of power in the plate circuit. This capability was essential in the design of audio output amplifiers that delivered substantial amounts of power when driven by the output of a low-power, voltage amplifier stage.

Note that by using the term *audio output amplifier*, we are rejecting the notion that the triode "amplifies power." Tubes are not technically power amplifiers. They merely control the power that is available in the plate circuit power supply.

The triode, with its directly heated filament, gained wide acceptance in radio receiver and transmitter design. The filament was very efficient in power consumption and well-fitted for early radio receivers that were battery-operated.

As AC power's popularity increased, pressure arose to power filaments from a common AC supply. The downside was that AC applied directly to filaments produced an annoying hum. Converting the AC to DC for the filament supply complicated the design and increased costs.

The solution was to replace the filament with an indirectly heated metal *cathode* as the emitter of electrons. Though less efficient, it solved the common supply and hum problems. Supplying power from AC mains readily overcame the efficiency issue.

The figure below shows the indirectly heated cathode and its schematic symbol.

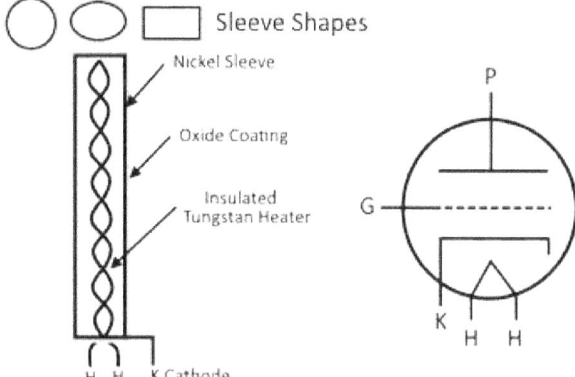

The cathode was basically a metal sleeve, usually nickel coated with emissive oxides, slipped over a tungsten heater. The sleeve's cross-sectional shape was round, oval, or rectangular, depending on the tube's application. By covering the tungsten wire with an insulating coating, it could be twisted, as shown, without making electrical contact with itself or the cathode. Twisting the heater wire helped to eliminate further induced AC hum.

Another important advantage of the indirectly heated cathode was that it allowed multiple triodes to independently operate while connected to a common AC heater supply source. Designers could connect heaters in series or parallel and remain independent of circuits associated with the cathodes.

The grid, or more properly, the *control grid*, consisted of fine wire wound in groves of supporting rods surrounding the cathode. See the figure below.

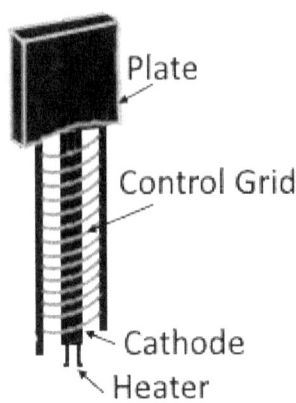

The size of the grid, along with the gauge of wire and its spacing, were all factors that affected a tube's characteristics. For receiving tubes, the dimensions of the tube elements were generally small, 1 to 2 cm in width, for example. For power triodes, electrode sizes were larger, especially the plate, as it had to dissipate considerable heat. With rectifier diodes, designers added fins to the plate to increase heat dissipation and output capability.

We can describe the operation of the triode using a set of three *characteristic curves*, each being a plot of three principal measurements: control grid voltage, plate voltage, and plate current. Each consists of a set of curves plotting two measurements for stepped values of the third measurement.

The most common characteristic curve is plate current versus plate voltage, which we measure by the steps of grid voltage. Below is the average plate characteristic for the 12AX7, a modern triode that we use later in our SET amplifier.

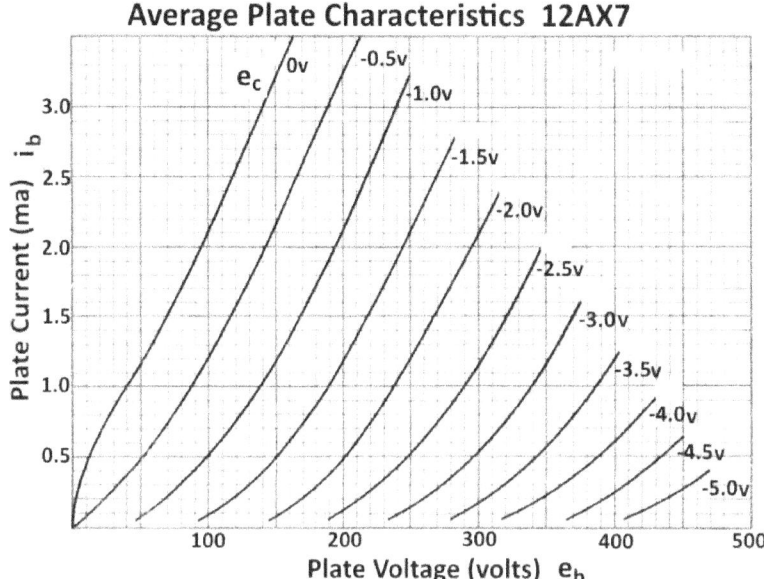

At a given grid voltage, the plate characteristic looks very much like the diode plate characteristic; that is, it steadily increases current with increasing plate voltage. Plots shift to the right as the grid voltage increases negatively. We expect this as the negative grid voltage alters the space charge in such a way that greater plate voltage is needed to offset its retarding effect on electrons leaving the cathode.

Next, we explore how the triode amplifier works by placing it in the circuit shown below.

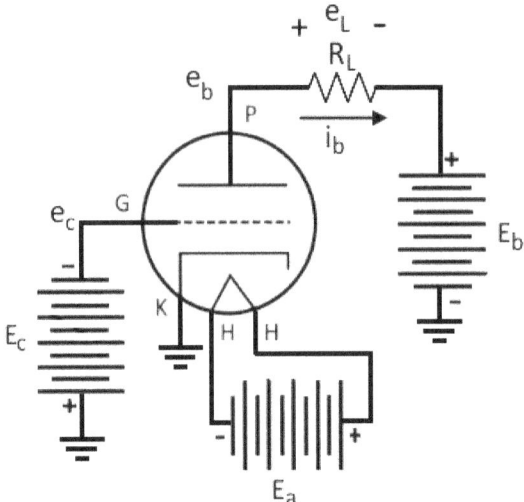

In this basic amplifier design, we ground the cathode. We call this circuit configuration a *common cathode*. Other configurations include the *common grid,* where we ground the control grid and the *common plate* or *cathode follower,* where we connect the plate to the plate supply.

A common cathode amplifier has a load element in series with the plate supply voltage. We apply the signal input to the control grid and take the amplified signal output across a load element.

In an audio voltage amplifier, the load is the resistor R_L. In audio output amplifiers, the load is a transformer and, indirectly, a speaker. In RF amplifiers, the load is a parallel inductor-capacitor (tank) tuned circuit.

To analyze this voltage amplifier, we first assume that the tube's plate characteristics (e_b vs. i_b) and supply voltages (E_a, E_b, and E_c) are known. The problem, then, is finding the plate current (i_b), plate voltage (e_b), and voltage across the load resistor (e_L).

The choice of "a," "b," and "c" to designate tube voltages is historic. The filament supply was originally from an "A" battery, the plate supply, a "B" battery, and the control grid supply, a "C" battery. When AC-powered devices came along, the A supply came from a low-voltage AC source, usually a transformer. The C supply came from a low-voltage DC source in the power supply or, more often, by way of *self-bias,* which we discuss later. The B plate supply also came from the power supply as a high-voltage DC source. Over time, the plate supply retained the letter designation, becoming the "B+" voltage source. We use "B+" to designate the plate supply source for our SET amplifier.

At this point, we need a little basic algebra to proceed. We first use Kirchhoff's Law from circuit theory to write the equation for the loop voltage in our triode circuit.

$$e_b - R_L \cdot i_b - E_b = 0$$

This equation basically means that we start with voltage e_b. As we move around the circuit, we encounter two voltage drops, $R_L \cdot i_b$ and E_b, reducing the voltage to zero.

Solving for i_b, we have

$$i_b = (-1/R_L) \cdot e_b + (E_b/R_L)$$

From algebra, we know this is a straight line with slope $(-1/R_L)$ and i_b axis intercept (E_b/R_L). We also know from the plate characteristics that i_b is a function of e_c and e_b, which we can write as $i_b = f(e_c, e_b)$, where "f" is a function known graphically via the plate characteristic curve.

What we have now are two equations in two unknowns.

$$i_b = (-1/R_L) \cdot e_b + (E_b/R_L) \quad (3\text{-}1)$$

$$i_b = f(e_c, e_b) \quad (3\text{-}2)$$

Since Eqn. 3-2 is graphical, we must pursue a graphical solution. Such a solution consists of plotting Eqn. 3-1 on the characteristic curves representing Eqn. 3-2 and then looking for the point of intersection between the curves.

To illustrate, let's assume a plate supply voltage of 300 volts and a load resistor of 220 kΩ. See the resulting plot below.

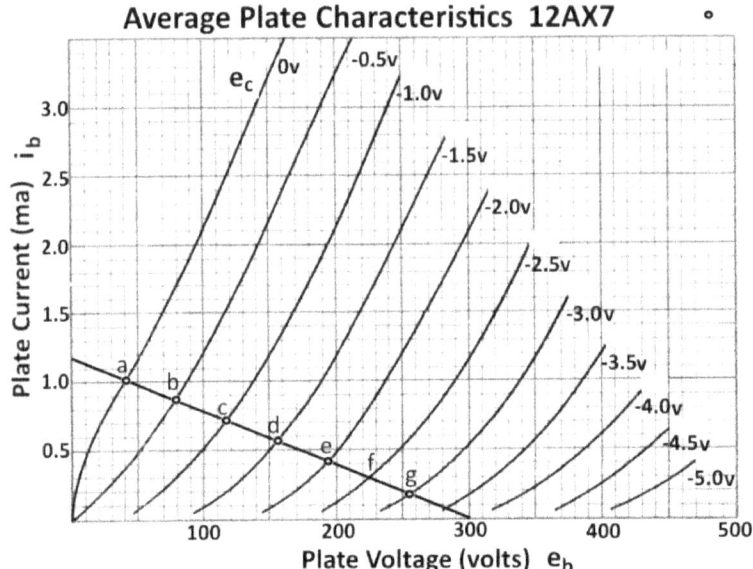

A straightforward way to plot Eqn. 3-1 is to mark two points that satisfy the equation and draw a line between them. Two such points are when $e_b = 0$ and when $i_b = 0$. Solving for i_b in the first case gives $E_b/R_L = 300/220,000$ or 1.36 ma. The left end of the line is then at point (0 volts, 1.36 ma).

With $i_b = 0$, we get $e_b = E_b = 300$ volts. The right end of the line is then (300 volts, 0 ma). After plotting these two points and drawing a line between them, we have the *load line* shown in the figure above.

The *static* or *quiescent* solution depends on which value of e_c we choose. Here are voltage amplifier design considerations to keep in mind when selecting a DC operating point:

1. **Keep $e_c \leq 0$** – Problems arise when we drive e_c into the positive region. The control grid begins to draw current supplied by the grid drive circuit, which is not possible with the capacitor coupling that we commonly use in voltage amplifier design. Therefore, we must choose the operating point so that the largest signal peak does not drive the control grid voltage e_c positive.

2. **Keep $e_c \geq$ cutoff voltage** – If e_c becomes too negative, plate current drops to zero and can be no less. The result is a flattening of the signal response whenever e_c is less (more negative) than the cutoff voltage. We call this signal *clipping* and must avoid it in design work.

3. **Choose an operating region where e_c steps show even spacing** - When a signal (a sinewave, for instance) is impressed on e_c, the operating point moves to values of e_c up and down the load line. Linearity and distortion remain good so long as the e_c curve steps along the load line are equally spaced. As the spacing becomes unequal, the degree of signal amplification varies, and the shape of the output signal is no longer an exact copy of the original, thus introducing non-linearity and distortion. Non-linearity results in distortion of various kinds. See Appendix C for more information.

For example, consider the 6J5 characteristics below.

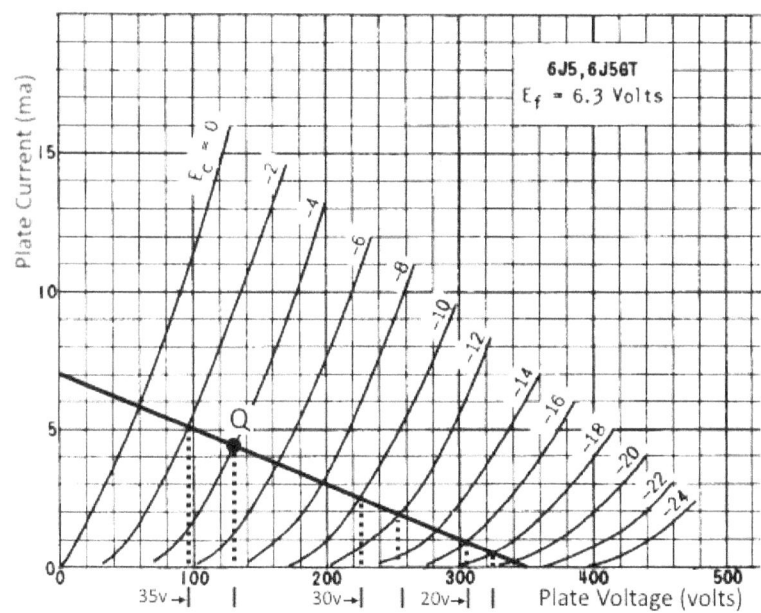

Note how the spacing between plate curves varies across the control grid voltage range. Toward the left end, spacing in plate voltage is 35 vdc, while on the right end, it shrinks to 20 vdc. To avoid non-linearity, we would do well to restrict voltage changes to ±4 vdc about the DC operating point Q. So long as the sinewave impressed on the control grid does not exceed ±4 vdc, we meet all three design considerations and minimize distortion.

Now, let's look at a more modern tube designed for low distortion, the 12AX7. First, we choose e_c = -1 vdc.

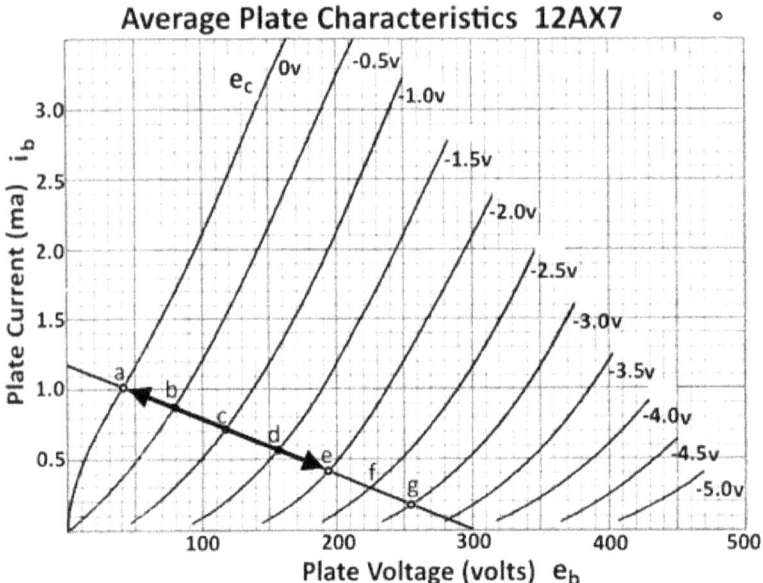

If we limit the signal input to a sine wave of 2 volts peak-to-peak (vpp) or less, the e_c spacing is nearly equal, and the resulting plate voltage swing linear. This DC operating point clearly meets all three design considerations.

Next, we look at what happens if we move the DC operating point to point e.

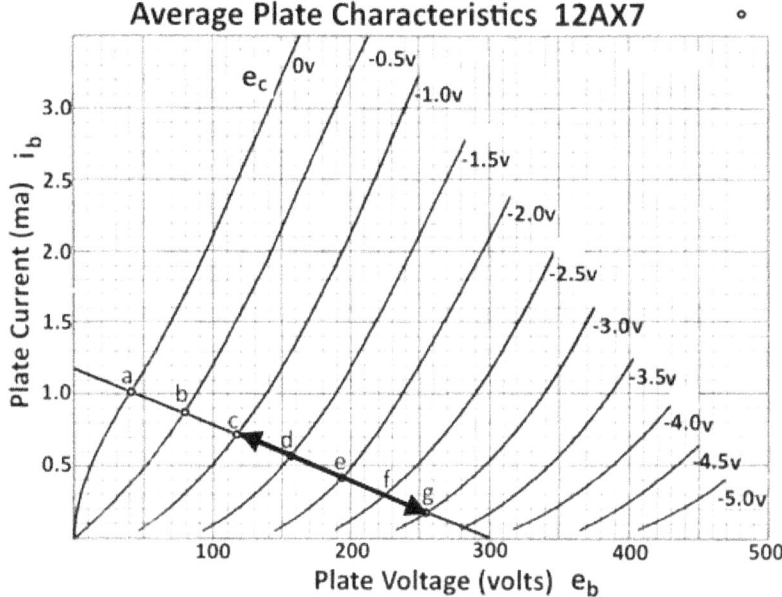

For a 2 vpp sinewave, the e_c step spacing above and below point e changes noticeably, which increases non-linearity and distortion, failing design consideration 3 above.

Lastly, we move the DC operating point to g.

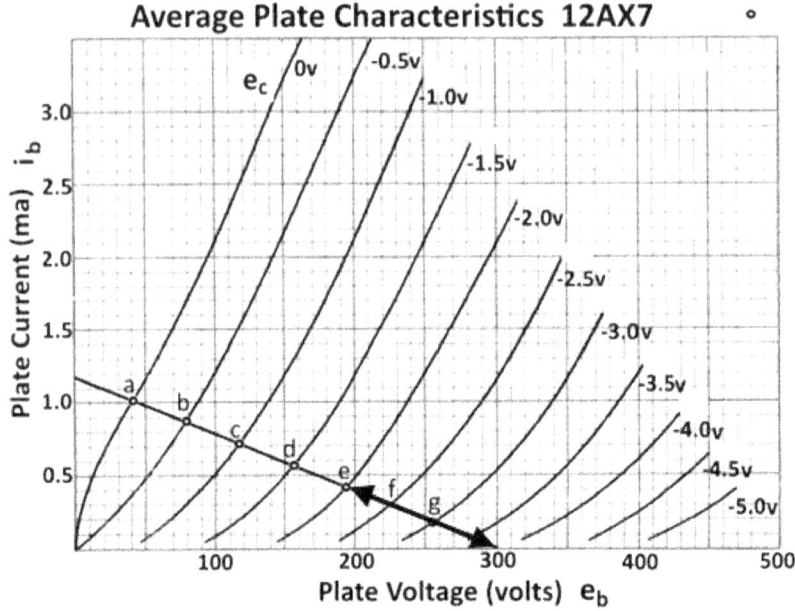

Not only does plate curve spacing vary in this region, but the 2 vpp sinewave drives the plate voltage into the 12AX7's cutoff region. So, we violate both considerations 2 and 3!

Settling on point c (e_c = -1 vdc), we can calculate voltage gain at Q.

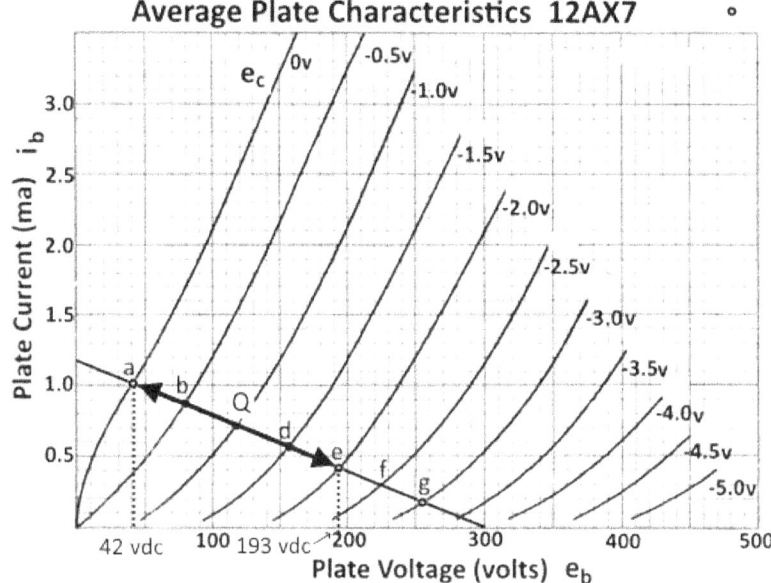

The difference in plate voltage above and below Q is (193 − 42) or 151 vdc. The variation of e_c is 2 vdc, so the voltage gain A_v = 151/2 ≈ 76.

The load line method of analysis used above is excellent for audio frequency (AF) amplifiers where signal levels are in the voltage range of a few control grid voltage steps. However, for much smaller signals (a few millivolts or less), such as those encountered in RF (radio frequency) amplifiers, we must use a different approach.

For *small signal applications*, we represent the triode using an equivalent circuit consisting of a voltage-controlled voltage source and a series resistance, as shown below.

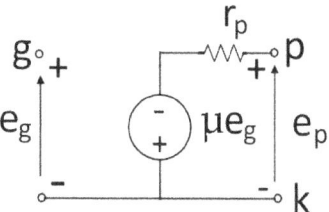

The parameters are as follows:

e_g – Small signal control grid voltage
e_p – Small signal plate voltage
μ - tube's amplification factor
r_p – tube's plate resistance

Tube manuals usually give values for μ and r_p as well as a recommended operating point where we can expect these parameters to have the given values.

(Note: In the discussions that follow, we use lower-case "e" and "i" to represent audio signal voltages and currents, respectively. Similarly, we use upper case "E" and "I" to represent DC voltages and currents.)

To calculate voltage gain A_V for the small signal equivalent, we use the circuit below.

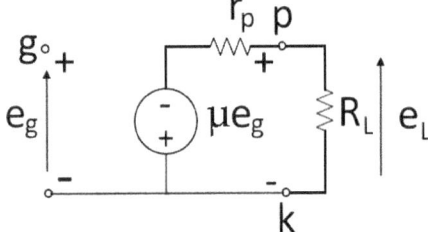

Voltage e_L is simply by the voltage divider rule equal to

$$e_L = -\frac{R_L}{R_L + r_p} \cdot \mu e_g$$

And

$$A_V = \frac{e_L}{e_g} = -\frac{\mu R_L}{R_L + r_p}$$

We should emphasize that the small signal equivalent circuit approach is approximate and only valid for values of ec in the microvolt to a few millivolt signal range. Also, the parameters vary with operating point, signal amplitude, and signal frequency. Thus, it is inappropriate to use the small signal approach when designing a SET voltage amplifier stage where the input voltage is about 0.5 to 1.0 vpp.

In this chapter, we developed the theory of voltage amplifiers.

In the next chapter, we apply the theory to practical voltage amplifiers like the one we use in our SET amplifier.

Chapter 4 – The Voltage Amplifier Triode

Thus far, we have discussed triode voltage amplifier circuitry in the theoretical realm. In a practical amplifier, we supply the heater voltage (E_a) and plate voltage (E_b or B+) externally with the power supply. We supply the control grid bias voltage (E_c) either externally with the power supply or internally within the amplifier circuit. Below, we describe these three ways to supply E_c: fixed bias, grid-leak bias, and self-bias.

Fixed-bias

The figure below shows a typical triode amplifier with *fixed bias* where we supply the control grid bias externally with the power supply.

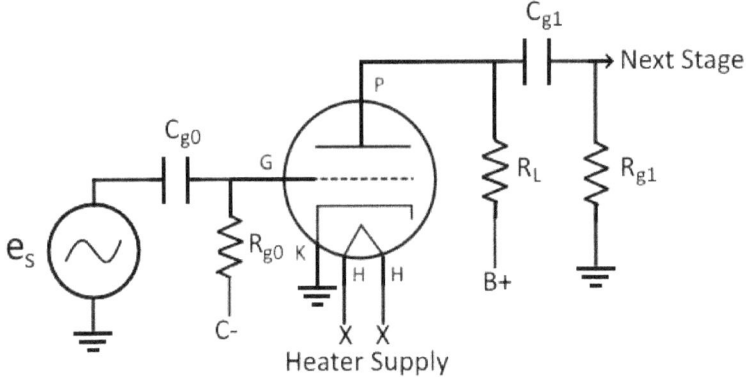

Here are the components and their function.

e_s – The sine wave signal generator.

C_{g0} – Input coupling capacitor that isolates the triode control grid from the signal source. We assume the capacitor is a short circuit at signal frequencies and an open circuit for DC voltage.

R_{g0} – Control grid input resistor that is sufficiently large so as not to overload the signal source yet not so large as to allow grid leakage current* to affect the grid bias and alter the operating point. We assume that in the absence of a signal, no DC current flows in this resistor and the no signal control grid voltage is that of the C- supply.

C- – The negative DC voltage source in the power supply that sets the DC operating point of the triode.

Heater Supply – The AC voltage source in the power supply that powers the tube's heater.

R_L – The plate load resistor.

C_{g1} – Output coupling capacitor that isolates the triode plate from the next stage. We assume the capacitor is a short circuit at signal frequencies and an open circuit for DC voltage.

R_{g1} – Control grid resistor for the following stage. We must include R_{g1} in the design because, at signal frequencies, capacitor C_{g1} is a short circuit placing R_{g1} in parallel with R_L. The additional load alters the load line (making it steeper), thereby reducing the voltage gain of the stage.

B+ – The plate supply that is a short circuit at signal frequencies and an open circuit (obviously) to DC voltage.

* Grid leakage current results from the flow (leakage) to the DC ground of a small number of electrons captured by the control grid. If the return path for this current is insufficient (R_{g0} too large), electrons build up on the grid, producing a bias voltage that drives the grid more negatively and alters the desired DC operating point. Control grid resistor values typically range from a hundred thousand Ω to a megΩ.

Under no signal conditions, the circuit establishes a DC operating point in an identical manner to the basic triode described in the previous chapter; that is, we can adjust the value of voltage C- to achieve a desired DC operating point. When a signal is present, capacitor C_{g0} acts as a short circuit, and the control grid voltage develops across R_{g1} and follows the signal voltage above and below C-. In turn, the plate current and voltage vary along the load line in accordance with the signal voltage e_s, rising above and falling below their DC operating values. The output signal passed through C_{g1} to the next stage is an amplified and inverted copy of the input signal e_c.

One point worth noting is that the load resistance is no longer just R_L. Since capacitor Cg1 and the B+ supply are short circuits at signal frequencies, the next stage's control grid input resistor R_{g1} is effectively in parallel with R_L. Thus, at signal frequencies, the combined resistance decreases the plate load resistance, steepens the load line, and decreases the stage's voltage gain.

To keep the terminology straight, we refer to the load line as the DC load line due only to DC components in the tube's plate-cathode loop. The load line that additionally includes the effect of components present only at signal frequencies (like R_{g1}) is called the *AC load line*.

(Although the abbreviation AC is associated with 60 Hz mains power, we also use the term more generally to apply toward sinusoidal signals of audio frequencies, as in *AC load line*.)

A characteristic of the AC load line is that it always passes through the DC operation point. Because the parallel combination of R_L and R_{g1} is smaller than R_L, the effect is to steepen the DC load line and lower the voltage gain. We defer giving an example until we consider the self-bias circuit below.

The advantage of fixed bias is that, with the cathode grounded, we take full advantage of the plate supply (B+) voltage and eliminate any components in the cathode that can affect gain and frequency response.

We rarely see fixed-bias used in SET voltage amplifier design. It is common in the SET power output stage, where it provides maximum output power for a given B+ voltage.

The disadvantage of the fixed-bias configuration is twofold. First, the addition of the bias power supply adds to the complexity and cost of the amplifier's power supply. Second, and perhaps more important, fixed bias does not compensate for the aging of the tube.

As the tube ages, emission decreases while fixed-bias voltage remains the same. The result is a shift in the DC operating point and possible degradation in amplifier performance. These age-related issues are not a particular problem with a voltage amplifier, but they can be significant in the power output stage. In the latter case, we must check the plate current periodically and adjust the fixed-bias voltage to maintain the desired DC operating point.

Grid-Leak Bias

Grid-leak bias voltage takes advantage of the small grid leakage current present at the control grid. As it passes through a high-value control grid input resistor R_{g0}, it creates the negative voltage needed to set a DC operating point. See the figure below.

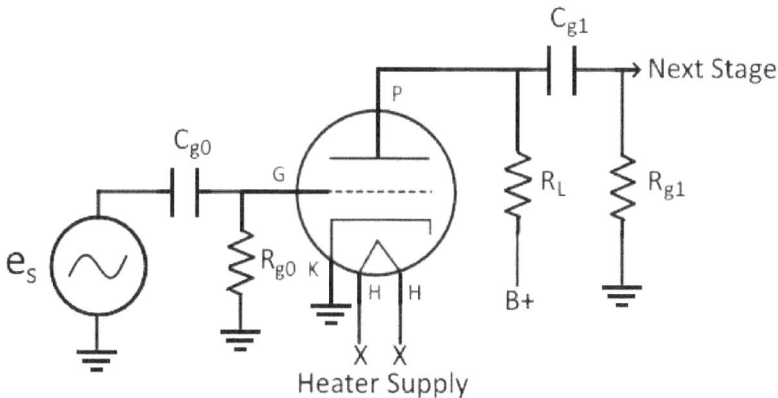

Typical values for grid-leak resistors range from 4.7 MΩ to 10 MΩ. Not all tubes are suitable for grid-leak bias. Typically, they are the ones with small grid bias voltages in the range of -0.1 to -0.4 volts.

As with fixed bias, grid-leak bias designs have the advantages of a grounded cathode but make no provision for tube aging. Grid bias is not suitable for a single-ended audio power output stage. Still, we occasionally see it in the voltage amplifier stage of inexpensive radios, where minimizing component count and cost override considerations in design.

Self-bias or Cathode-Bias

Lastly, we come to the most frequently used grid bias method in voltage amplifiers, *self-bias* (sometimes referred to as *cathode-bias*). See the figure below.

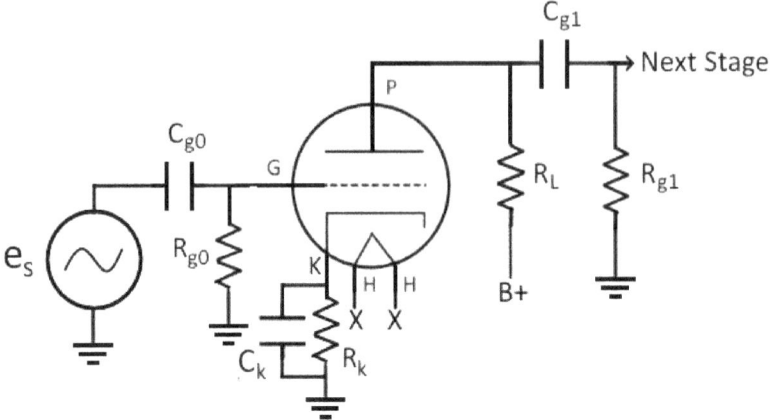

Because the control grid current is essentially zero under normal operating conditions, the cathode current is equal to the plate current. We take advantage of this by inserting resistor R_k between the cathode and ground. We choose the value so that the cathode current produces a voltage drop equal to the desired control grid bias. Because the voltage at the cathode side of R_k is positive with respect to the DC ground, the control grid voltage with respect to the cathode is negative, as required for control grid bias.

Since R_k is in series with the cathode-plate circuit, it reduces the effective plate voltage. In voltage amplifier designs, the cathode voltage is usually much less than the B+ voltage, and we can neglect it. In power output stages, this is not generally the case, and we must consider it.

Consider the voltage amplifier given in the circuit below.

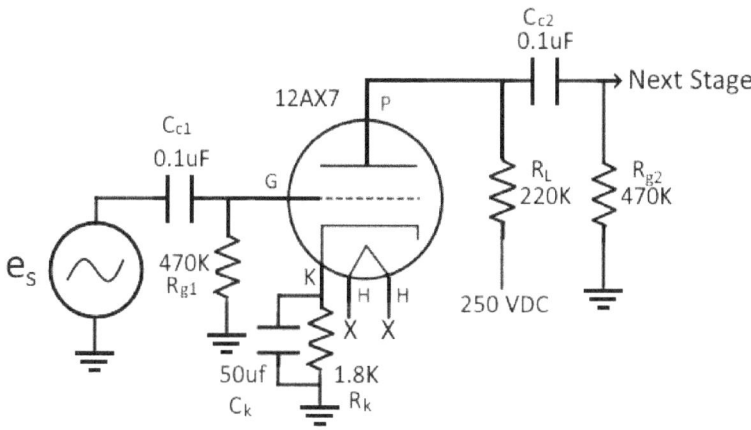

R_L is 220KΩ, and R_k has a much smaller value of 1.8KΩ. In fact, R_k is within the 5% tolerance range of R_L, meaning that for all practical purposes, we can ignore it in the DC load line calculation. The plate current is about 1 ma, meaning that the voltage drop across R_k is ~2 vdc, which is much, much (<<) less than the B+ voltage, 250 vdc. We can ignore it when drawing the voltage amplifier's load line.

In the figure below, we have neglected R_k when drawing the DC load line.

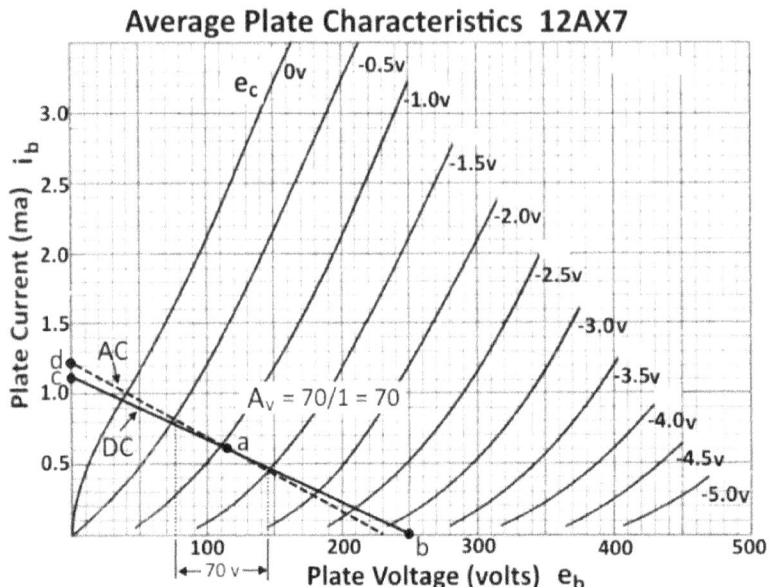

Point "b" of the DC load line is the plate source voltage 250 volts. Point "c" is the supply voltage divided by the plate resistor R_L or $250/220 \times 10^3$ = 1.1 ma. The DC load line passes between these two points.

To determine R_k, we begin by noting that we want the operating point "a" to be where the DC load line crosses the control grid voltage curve labeled "-1.0 volts". The current at "a" is ~0.6 ma. To size the resistor, divide 1.0 volts by 0.6 mA to get R_{g0} = 1.7KΩ. We would use R_k = 1.8 KΩ as the closest 5% value. The DC operating point is at "a" where $e_c \approx -1.8 \cdot 10^3 \cdot 0.6 \cdot 10^{-3}$ = -0.1 vdc.

To draw the AC load line, we know that one point is the DC operating point "a." For AC signals, we bypass R_k to ground with C_k so we can ignore it to determine the AC load line. To get the other point, we find the parallel combination of 220K//470K, which is $220 \times 470/(220+470)$ = 150KΩ. Next, we note that the DC operating plate voltage is ~113 volts and divide by this value. We get 113/150K = 0.75 ma. We add this to the operating point current of 0.6 ma, and we get 1.35 ma, which is point "d," the AC load line intercept with plate current axis. Finally, we draw the AC load line between "d" and "a," carrying the line onto the plate voltage axis.

To calculate the voltage gain, we use the AC load line and choose, for convenience, a 1 vpp variation in e_c that is the denominator of the A_v calculation. As shown in the figure, the plate voltage difference at e_c equal -0.6 volt and -1.6-volt crossings is ~70 vpp. The ratio of this to the input voltage 1 vpp is the voltage gain or $A_v \approx 70 / 1 = 70$.

Now is a suitable time to point out why we used the approximate symbols "~" and "≈" in the text. Because of component tolerance and tube characteristic variation, most tube design calculations, at best, are approximate. Variations of 5% to 10% are not unusual, which is the impetus for our later breadboarding tube designs before committing to construction!

To assist in voltage amplifier design, tube manuals sometimes provide component values for parameters such as voltage gain A_v, output voltage (peak-to-peak), and B+ voltage. Below is the 12AX7 table from the 1966 RCA Receiving Tube Manual.

B+	R_p	R_g	R_k	C_k	C_c	E_o	A_v
180	0.1	0.1	1800	4.0	0.025	18	40
	0.1	0.22	2000	3.5	0.013	25	47
	0.1	0.47	2200	3.1	0.006	32	52
	0.22	0.22	3000	2.4	0.012	24	53
	0.22	0.47	3500	2.1	0.006	34	59
	0.22	1.0	3900	1.8	0.003	39	63
	0.47	0.47	5800	1.3	0.006	30	62
	0.47	1.0	6700	1.1	0.003	39	66
	0.47	2.2	7400	1.0	0.002	45	68
300	0.1	0.1	1300	4.6	0.027	43	45
	0.1	0.22	1500	4.0	0.013	57	52
	0.1	0.47	1700	3.6	0.006	66	57
	0.22	0.22	2200	3.0	0.013	54	59
	0.22	0.47	2800	2.3	0.006	69	65
	0.22	1.0	3100	2.1	0.003	79	68
	0.47	0.47	4300	1.6	0.006	62	69
	0.47	1.0	5200	1.3	0.003	77	73
	0.47	2.2	5900	1.1	0.002	92	75

3AV6
4AV6
6AV6
6EU7*
12AV6
12AX7A/
ECC83*
20EZ7*
7025*

* One triode unit.   * Peak volts.

The headings refer to this circuit diagram.

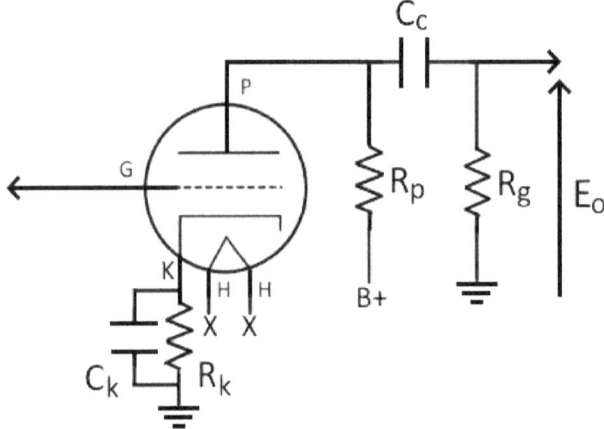

We use this chart later to design the voltage amplifier stage of our SET amplifier.

In the next chapter, we apply the theory so far discussed to the power output stage.

Chapter 5 – The Power Output Triode

As we have seen, achieving voltage gain is easy with a triode. But, to drive a speaker, it takes power. Specifically, we need a combination of low voltage and high current. For instance, to drive an 8-Ω speaker at 1 watt requires a current of nearly 1 amp peak-to-peak at only 8 volts peak-to-peak!

As we have found, vacuum tubes, by their very nature, are good voltage amplifiers, not power amplifiers. They merely control the power available from the B+ supply. The tube's power handling capability and inherent plate resistance limit control of available power.

We can find power triodes that have good power handling capability, well into 10s of watts. They, too, have low plate resistance compared to voltage amplifier tubes like the 12AX7. While in the 500 Ω to 1500 Ω range, their plate resistance is still too high to directly drive a speaker load in the 4 to 16 Ω range. With this ratio of plate-to-load resistance, the tube dissipates most of the available power with little power ever reaching the speaker.

The way around this is to use an audio output transformer like the one shown below.

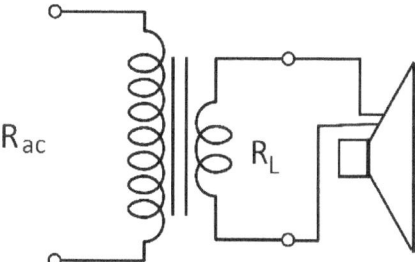

The audio transformer transfers sinusoidal signal power from its primary to its secondary with little energy loss. The equation below represents power transfer for a transformer.

$$e_p \cdot i_p = e_s \cdot i_s$$

e_p and i_p are the primary signal voltage and current, and e_s and i_s are the secondary signal voltage and current. The ratio of the transformer's primary to secondary winding turns *n* determines these values. That is,

$$\frac{e_p}{e_s} = \frac{i_s}{i_p} = n$$

For example, to reduce the speaker current of 1 amp peal-to-peak to a value within the capability of a typical power triode, we can use an audio transformer with turns ratio n of 25 to 1 (usually written 25:1). From the equation above, the primary current is $i_p = \frac{i_s}{n} = \frac{1.0}{25} = 40$ mA peak-to-peak, well within the capability of power triodes.

Rather than work in voltage and current ratios, it is simpler to think of the audio transformer coupling based on impedance. (See Appendix H for a discussion of "resistance" and "impedance.") To do this, we talk about matching the impedance of the speaker with the recommended load resistance of the power triode.

To find the impedance relationship to turns ratio, start with the voltage and current relationships above and rearrange them as shown below.

$$\frac{e_p}{e_s} = n \qquad \frac{i_p}{i_s} = \frac{1}{n}$$

Dividing the two equations assuming that $\frac{e_p}{i_p} = r_p$ (the primary impedance) and $\frac{e_s}{i_s} = r_s$ (the secondary impedance), we have

$$\frac{r_p}{r_s} = n^2$$
$$r_p = n^2 \cdot r_s$$

A transformer's primary impedance then is equal to its secondary load impedance multiplied by the square of the turns-ratio or n^2.

For example, suppose the turns ratio n of an audio transformer is 25; then an 8-Ω secondary impedance load (such as a speaker) appears as 5000-Ω impedance.

[8 · (25)² = 8 · 625 = 5000] load across the primary. The speaker load appears to be in the same order of magnitude as the plate load resistance, and hence, we can expect a more acceptable transfer of power.

Manufacturers specify audio transformers using primary and secondary impedance. For instance, we would specify the audio output transformer described above as matching "a 5000-Ω impedance to an 8-Ω speaker". Note that the same transformer would match a 2500-Ω load to a 4-Ω speaker or a 10,000-Ω load to a 16-Ω speaker!

It is customary practice to construct an audio transformer secondary with taps so that it can match multiple speaker impedances. For instance, we might find a 5000-Ω audio output transformer with a common and 4-Ω, 8-Ω, and 16-Ω secondary connections.

Transformer and speaker properties vary with frequency. For instance, the specified impedance of a speaker is a nominal value, sometimes measured at a specified frequency or range of frequencies. In fact, speaker impedance can vary widely, often peaking significantly at a resonant frequency. We should also note that the speaker enclosure affects its load impedance. All this is to say that the impedance of a 4-Ω speaker is not like a 4-Ω resistor except at 0 Hz (DC)!

The figure below shows the impedance of an inexpensive 4-Ω speaker plotted from 10 to 10 KHz.

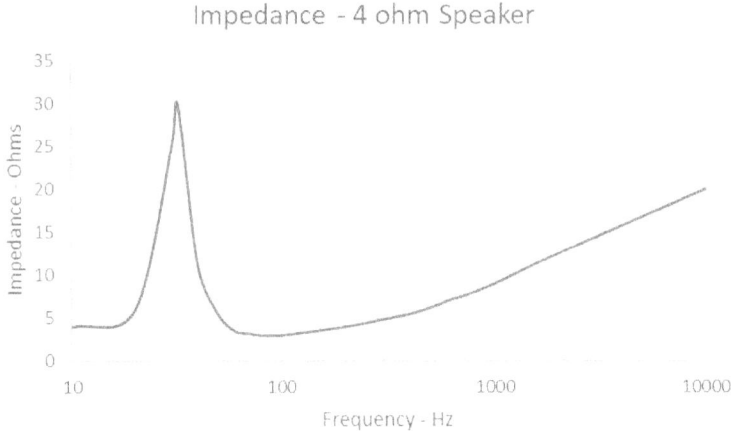

A resonant peak in frequency occurs below 100 Hz with a gradual increase above 100 Hz. The nominal frequency of 3-4 Ω only occurs for a small range of frequencies around 100 Hz and below 20 Hz.

The loudness of sound increases at frequencies near resonance. This results in an effect called *false bass*. Interestingly, radio manufacturers take advantage of false bass with small speakers to enhance male voices and certain musical passages. Otherwise, a 4-inch speaker in a table radio would sound "tinny" without the false bass content. The low-frequency resonance also accounts for the "mellow" sound attributed to old radios, particularly console models with 8- and 12-inch speakers.

From a high-fidelity point-of-view, a speaker's resonance peak is highly undesirable as it muddles low-frequency sounds, making them indistinguishable from one another. We place high-quality speakers with a very low-frequency resonance in the audible range in an enclosure that reduces the resonance effect.

An ideal audio transformer would present a flat frequency response across all audio frequencies. The frequency response of real audio transformers usually falls off at low frequencies, may peak slightly at mid to high frequencies, and then falls off rapidly at still higher frequencies. Below is the frequency response of a commercial transformer, the Hammond 125CSE.

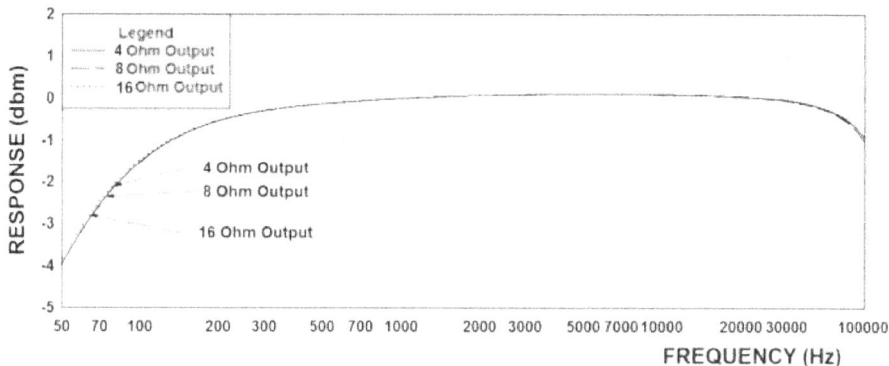

While the 125CSE's mid and high-frequency response is quite good, the low-frequency response falls off rapidly below 200 Hz.

In addition to limited frequency response, real transformers have limited power handling capability. For instance, the Hammond 125CSE transformer rating is 8 watts with a maximum plate current of 60 ma.

Frequency response and power handling capability come with a price. For power triodes, transformers with a power rating of 3-8 watts and reasonably flat frequency response of 150 to 15,000 Hz can carry a price tag of $50 or more. To get a transformer with the same power rating but with a high-fidelity frequency response of 20 to 20,000 Hz, the price tag easily doubles to $100 or more. A 20-watt high-fidelity transformer could cost over $200!

Now that we know how to handle the impedance matching problem, we need to specify the power triode's optimum load impedance R_{ac}. Given this value for R_{ac}, we can then select an audio transformer to match our speaker impedance.

The obvious choice is to make R_{ac} equal to the power triode's plate resistance. This choice maximizes power transfer from the power triode to the speaker.

To confirm this, assume that the power triode is a linear device. We can then represent the output circuit of the power triode as linear, voltage-controlled source e_p with internal resistance r_p in series with external load R_{ac}. The tube's control grid voltage e_c controls the magnitude of AC signal e_p. See the diagram below.

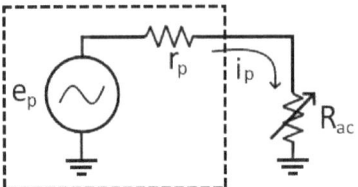

The formula below gives the power transferred.

$$P = i_p^2 \cdot R_{ac} = (\frac{e_p}{r_p + R_{ac}})^2 \cdot R_{ac}$$

Examining the equation, we see that power delivered to R_{ac} approaches zero as $R_{AC} \rightarrow 0$ and as $R_{AC} \rightarrow \infty$ which suggests that R_{ac} must reach a maximum at some intermediate point. With the help of a little "higher math," Calculus to be exact, we can show that the maximum power transfer occurs when $R_{ac} = r_p$. See Appendix A for proof.

Let's try $R_{ac} = r_p$ in a real case and see where it takes us. Consider the 2A3 power triode plate characteristics below. (Next page, please.)

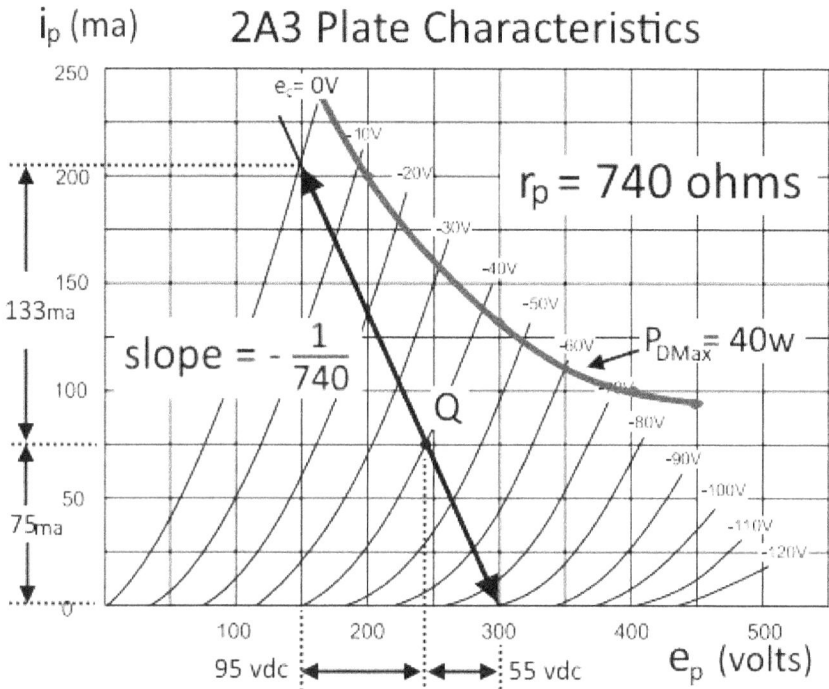

Assume the DC operating point is at Q (75 ma, 245 volts). (We often use "Q" to denote the DC operating point. It stands for "quiescent".) We find the plate resistance at this point by inverting the slope of the plate curve at Q. Estimating from the graph above, we have (262-225)/(0.100-0.050) = 740 Ω. If we assume $R_{ac} = r_p = 740$ Ω, then the slope of the load line is the negative inverse of R_{ac} or -1/740.

To draw the load line, we know that one point is Q. Another point is where the load line crosses the e_p axis (the e_p intercept). This is $E_Q - I_Q$/(slope of load line) or 250 - .075/(-1/740) = 300 volts. We draw the AC load line from the I_p intercept at 300 vdc through Q.

If we apply an 80 vpp input signal to the control grid, the operating point shifts along the AC load line, extending to the arrow points shown. We calculate the power output estimate using the formula below.

$$P_{out} \approx \frac{V_{pp} \cdot I_{pp}}{8} = \frac{150 \cdot .208}{8} = 3.9 \; watts$$

It is not a bad power output for a single-ended power triode, but there is something more we must consider.

Note that along the load line, the spacing of plate curves varies, becoming more compressed for increasingly negative values of e_c. To get some idea of how much variation, consider this. From e_c = -40 vdc to 0 vdc, the plate voltage varies to 95 vdc, while for e_c = -40 to -80 vdc, it varies to 55 vdc! Thus, the 2A3 is exceedingly non-linear.

While power transfer for R_{ac} = r_p is quite good, the bottom half of an input sinusoid would experience much less amplification than the top half! The slightly flattened and misshaped output sinusoid would significantly affect sound quality.

Making R_{ac} < r_p would steepen the load line even more and worsen the spacing variation in plate curves. Making R_{ac} > r_p would tend to even out the spacing of plate curves and lessen distortion. However, this would be at the expense of power transfer since we are moving away from R_{ac} = r_p. But how much greater do we make R_{ac}? Let's do a little more analysis.

We should note here that the significant non-linearity we have just identified calls into question the original assumption that the power triode was a linear device. So, let's back up a bit and be more careful with our assumptions!

First, we should establish some criteria that consider the non-linear plate characteristics of the triode. (1) We must not drive e_c positive; doing so with capacitive coupling would clip the audio input to the power output stage and cause unacceptable distortion. (2) We must avoid the region of low plate current in the *knee of the plate curve*. In this region, r_p changes radically, eventually driving the tube to cut off and clipping the audio output. In the 2A3 example above, we would avoid the i_p region below about 25 ma. (3) We must place the DC operating point to produce an equal swing of e_c between the limits established by criteria 1 and 2.

Based on these criteria, we establish these limits on drawing the load line: (1) The upper limit is the plate curve for which e_c is zero, and (2) The lower limit is the plate curve where the most negative e_c plate curve intersects the line i_p = I_{min}. The graph below shows an idealized load line subject to these limits.

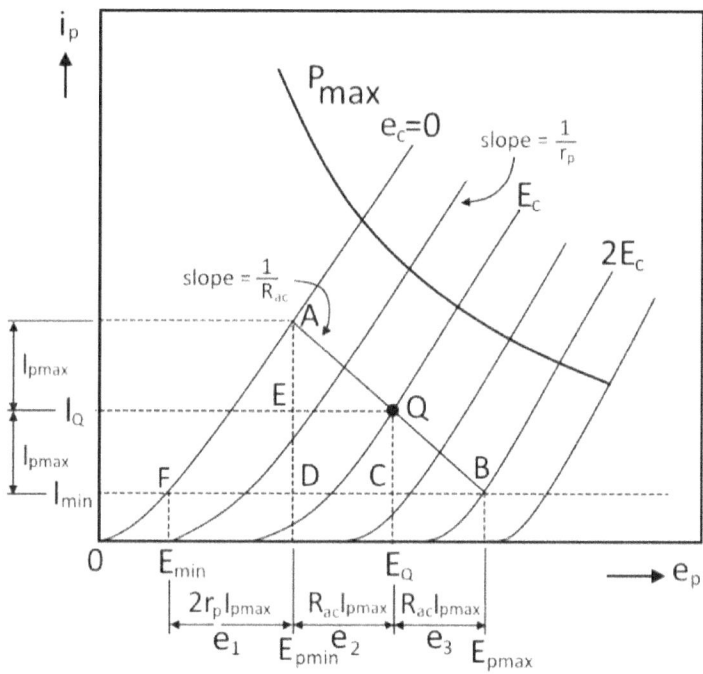

Using a little geometry and similar triangles, we can determine values for lengths e_1, e_2, and e_3 in the figure. See Appendix B for details. From e_1 and e_2, we have

$$E_Q - E_{min} = I_{pmax} \cdot (2r_p + R_{ac}).$$

From this, we find

$$I_{pmax} = \frac{(E_Q - E_{min})}{(2r_p + R_{ac})}$$

Power output is

$$P_{out} = \frac{I_{pmax}^2}{2} \cdot R_{ac} = \frac{R_{ac} \cdot (E_Q - E_{min})^2}{2 \cdot (2r_p + R_{ac})^2}$$

Using calculus once again, we find that maximum power output occurs when $R_{ac} = 2r_p$. See Appendix B for the proof.

In the figure above, the control grid lines are equally spaced, so the gain across the load line is linear, and therefore the distortion is zero. In a real power triode, plate characteristic lines are not equally spaced, which gives rise to second-order harmonic distortion.

As already noted, increasing R_{ac} above r_p decreases distortion, and now we know that, given the specified criteria, the power peaks at $R_{ac} = 2r_p$. The figure below shows the experimental result of harmonic distortion and power output versus R_{ac} for a 6BQ5 power pentode with its screen connected to its plate. The result is typical of most power triodes.

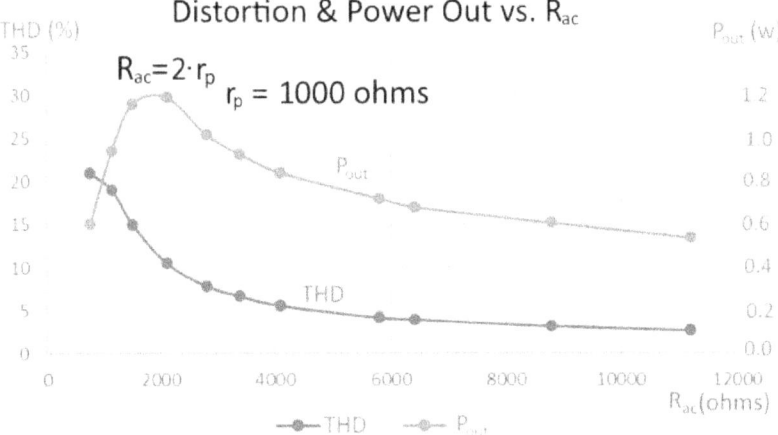

Note that, as expected, maximum power transfer occurs at $R_{ac} = 2 \cdot r_p$ or 2000 Ω, where harmonic distortion is still unacceptably high at 11%. As R_{ac} increases above $2 \cdot r_p$, distortion falls off, but so does power output.

The strategy then is to select a value of R_{ac} above $2 \cdot r_p$ that gives acceptable power output for a specified level of harmonic distortion. If we cannot achieve this with the tube at hand, then we look for a tube with higher power handling capability.

In actual practice, designing a triode power stage can be and often is an iterative process. We can't just plug some numbers into a formula and get an ideal design. One thing that would help would be a straightforward way to estimate harmonic distortion given a DC operating point, value for R_{ac}, and resulting AC load line.

Consider again the idealized triode plate characteristics shown below. (Next page, please.)

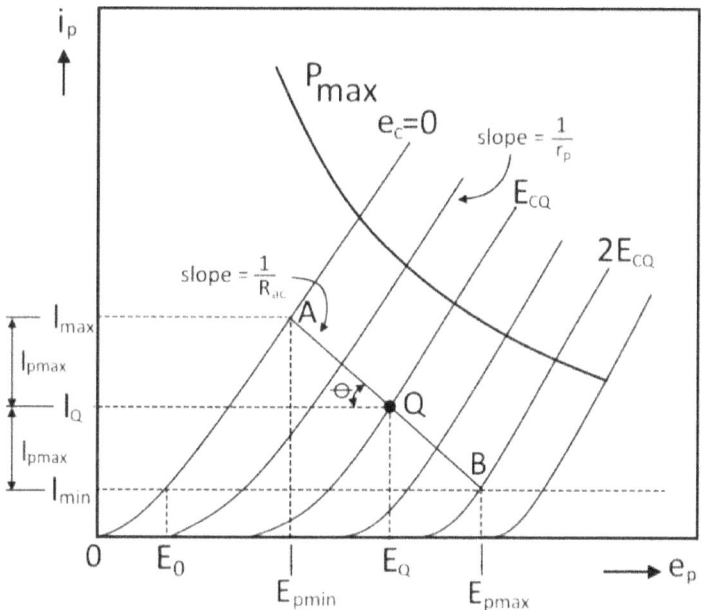

If we start with these assumptions:

1. The control grid voltage maximum is 0 volts.
2. Plate current is never less than the "knee" of the i_p vs. e_p curves, I_{min} in the figure.

3. The operating point Q is below the maximum plate dissipation of the tube, P_{max}, in the figure.

4. The minimum (most negative) value of control grid voltage is twice the value at the operating point.

5. Harmonic distortion is second order only.

We show in Appendix D that, given these assumptions, we have,

$$Triode\ Harmonic\ Distortion \approx \frac{\frac{AQ}{QB}-1}{2\left(\frac{AQ}{QB}+1\right)} \cdot 100\%$$

$$Power\ output \approx \frac{1}{8}(E_{pmax} - E_{pmin}) \cdot (I_{max} - I_{min})\ watts$$

The chart below shows a plot of harmonic distortion versus the ratio AQ/BQ.

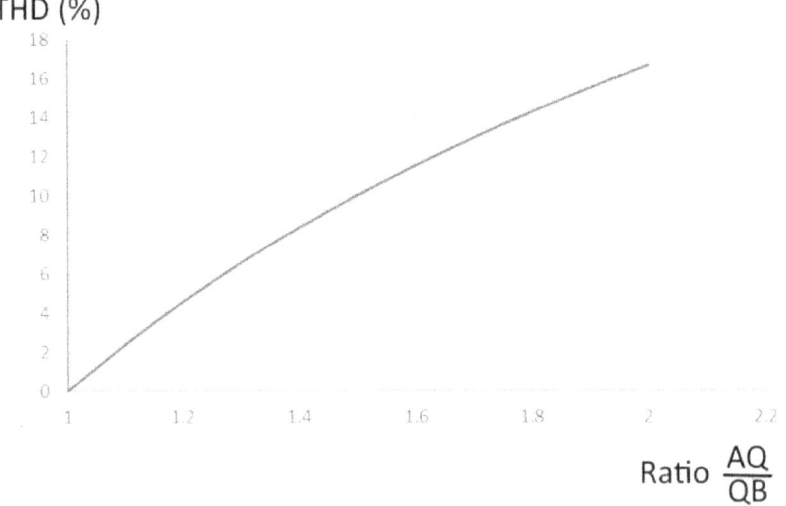

Ratio $\frac{AQ}{QB}$

At last, we have the tools we need to choose R_{ac}. Here are the steps.

1. Select the operating point (I_Q, E_Q). Often, tube manual specifications include a recommended operating point for the power triode. If this is not available, choose E_Q and calculate a trial value of I_Q as $E_Q/(4r_p)$, where we calculate r_p from the plate curve near the operating point. If the calculated (I_Q,

E_Q) is far from the assumed operating point, calculate a new r_p at the newly calculated (I_Q, E_Q) and redo the I_Q calculation. With each iteration, the DC operating point and r_p more closely match.

2. Using the chart above, determine the ratio of AQ/BQ for the maximum acceptable second-order distortion. For 4%, AQ/QB is about ~1.18.

3. Starting with R_{ac} = 2r_p, draw a load line and check the ratio of AQ to QB. Since the ratio is dimensionless, we can use any convenient measurement unit. A unit of millimeters works well for the tube's plate characteristics printed on a letter-size sheet of paper. Try increasing Rac values until the ratio of AQ to BQ is equal to or less than the designated amount, 1.18 in the example.

4. Calculate the power output with the selected R_{ac} using the formula

$$Power\ output \approx \frac{1}{8}(E_{pmax} - E_{pmin}) \cdot (I_{max} - I_{min})$$

If it is acceptable, then choose an audio output transformer that matches this impedance to the speaker impedance. Be sure we meet the current (based on I_Q) and power handling requirements. If the power output is less than acceptable, then try a power triode with a larger power handling capacity and repeat the procedure.

Below is an example of the method using a 2A3 power triode. Our goal is to get at least 3 watts with no more than 2% distortion. To prepare, print the 2A3's plate characteristic curves as large as possible on a letter size paper. Have a pencil and ruler handy. Using the distortion graph above, find that, for a 2% distortion, we need the ratio of AQ to QB to be 1.1 or less.

Step 1. For the DC operating point Q, we use the values suggested in the RCA Tube Manual, 60 mA at 250 volts.

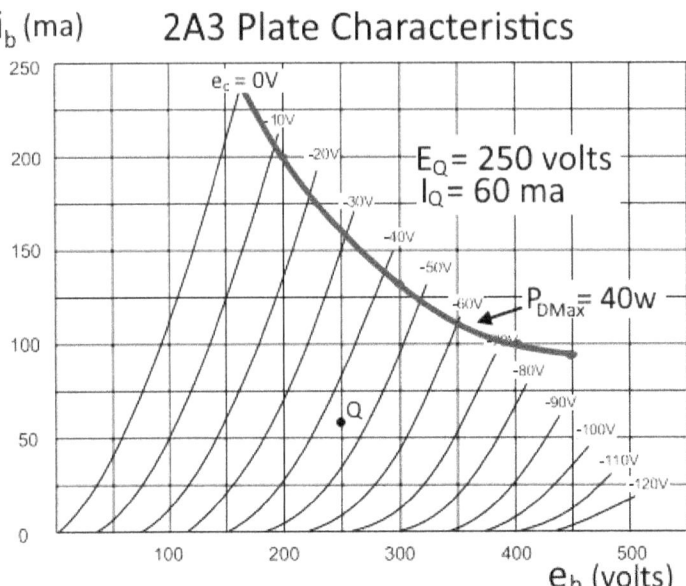

The bold curve represents the maximum plate dissipation permitted for the tube. That is, the value of $E_Q \cdot I_Q$ must be less than 40 watts or, graphically, the Q point must fall below the curve. Though it is not necessary, the AC load line should stay beneath the curve, as well.

Step 2. Calculate the r_p at the operating point. See the figure below. (Next page, please.)

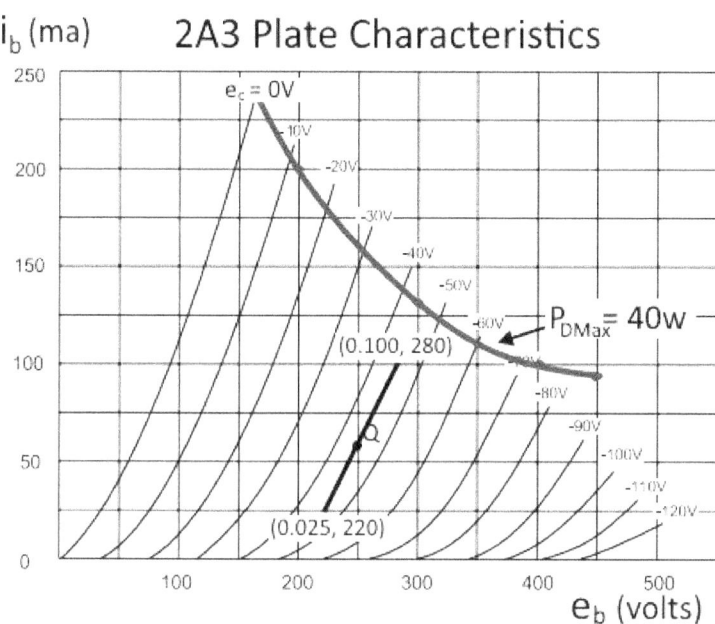

Draw a short line through Q, paralleling the plate curves and ending at convenient current and voltage lines. The latter makes reading values easier. Make a note of the endpoints: (0.100,280) and (0.025, 225). Calculate $r_p = \Delta V / \Delta I = (280 - 220) / (0.100 - 0.025) = 800\ \Omega$.

Step 3. Start with an AC load line with a slope of $-1/2r_p$ or $-1/1600$. For a quick way to draw an AC load line, start with the figure below.

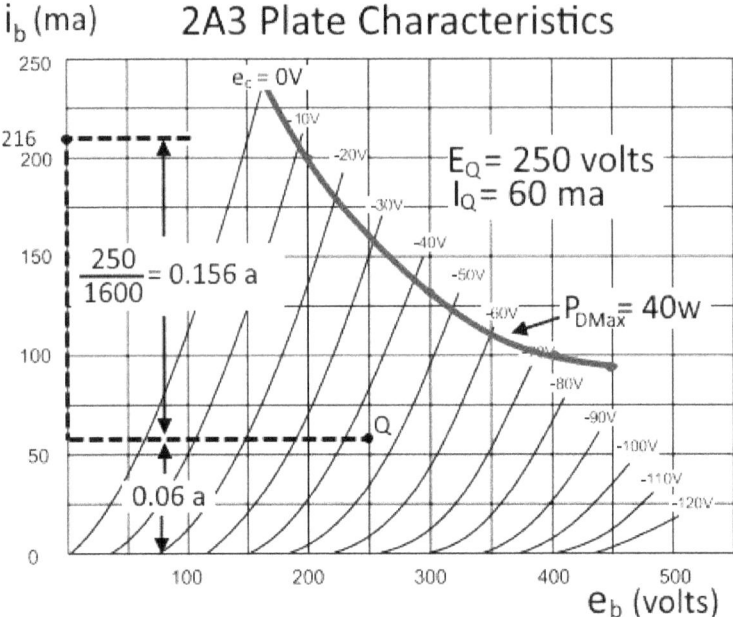

Divide E_Q by the load line resistance, in this case, 1600 Ω. We have 250 volts / 1600 Ω = 0.156 a. Add this to I_Q 0.156 + 0.060 = 0.216 a. This is the i_p axis intercept. Plot the point (0.216, 0) and draw the load line through point Q. See line 1 below.

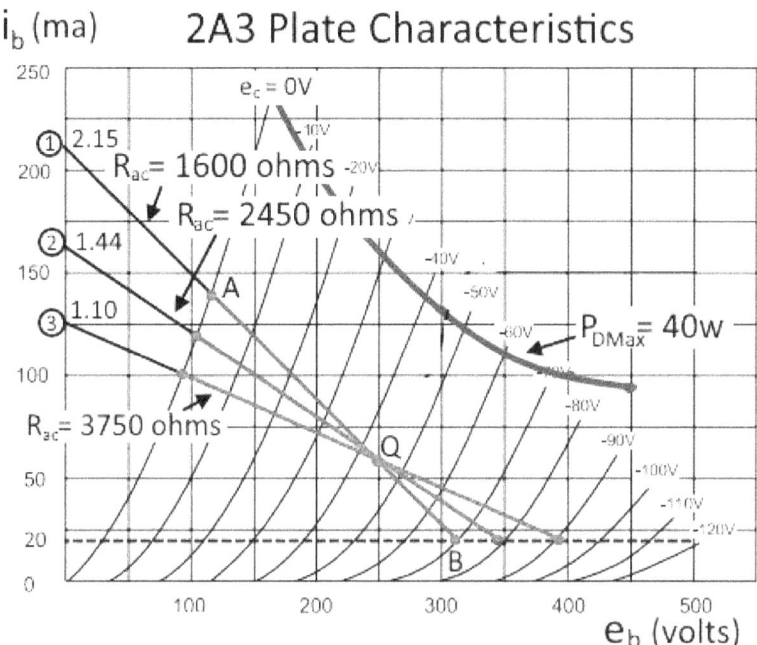

Note that the maximum value of e_c is 45 volts above E_Q. We must restrict e_c's value below E_Q to the same minimum value or -90 volts. But in any case, we must limit plate current at the knee of the plate characteristic, which is the value I_{min} in the figure.

Looking at the figure, the line segment of interest starts at the characteristic curve for $e_c = 0$ (point A) and ends at I_{min} (point B). Had the line extended to the -90-volt curve before reaching I_{min}, we would have ended it there instead.

Step 4. Measure the length of AQ and QB using a millimeter scale. From the figure above, we have AQ = 71 mm and QB = 33 mm. The ratio of AQ to QB is 71 / 33 = 2.3, considerably larger than our target of 1.1. Apply the 2.3 ratio to the distortion chart; the estimated distortion is 20%. We calculate power out using this equation.

$$P_{out} = \frac{(I_{max} - I_{min}) \cdot (E_{max} - E_{min})}{8}$$

$$P_{out} = \frac{(.140 - .020) \cdot (312 - 120)}{8} = 2.9 \; watts$$

The power out is good, but the distortion is unacceptably large!

Step 5. Using larger values of R_{ac}, we repeat steps 3 and 4 until the AQ to QB ratio drops to 1.1 or less.

Here are the results.

Line	AQ	QB	AQ/QB	THD	P_{out}	R_{ac}
1	71	33	2.3	20%	2.9 w	1600 Ω
2	65	45	1.44	9%	3.1 w	2400 Ω
3	64	58	1.1	2%	3.0 w	3750 Ω

Load line 3 with R_{ac} = 3750 Ω meets our requirements and does so with a power output of 3 watts.

As a practical matter, finding an audio output transformer with a primary impedance of 3750 Ω may not be possible other than as a custom manufacturer. In practice, an impedance selection within a few hundred Ω makes a minor difference. In this case, a 3500-Ω primary transformer with 4-, 8-, and 16-Ω secondary is readily available (Hammond #P-T1630SEA).

Power triodes tend to be rare and expensive. In the next chapter, we find that we can use more readily available power pentodes instead.

Chapter 6 – Triode Connected Pentodes

As a rule, quality power triodes are not readily available, and when they are, they generally carry hefty price tags. A more practical choice is a triode-connected power pentode. With its screen grid connected to its plate, the power pentode's plate characteristics are like that of a power triode.

A variety of power pentodes are readily available as both new-old-source (NOS) and newly manufactured, plus their cost is far more reasonable. Power output ratings range from 1 to several watts, making them acceptable for the design approach in the previous chapter.

Power pentodes include a screen grid and suppressor grid that increase tube performance, particularly in power handling capability. To triode connect a pentode, we tie the screen grid to the plate. See the figure below.

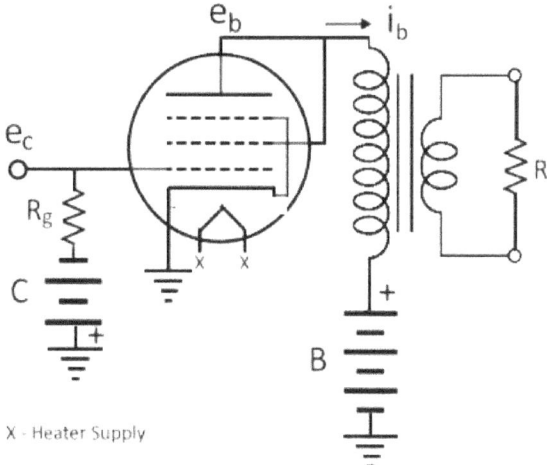

As in the figure, the suppressor grid and cathode often connect internally. If this is not the case, we connect the suppressor grid to the plate along with the screen grid.

As previously noted, the plate characteristics of a triode-connected pentode resemble very closely that of a triode. See the 6F6 triode-connected plate characteristics below.

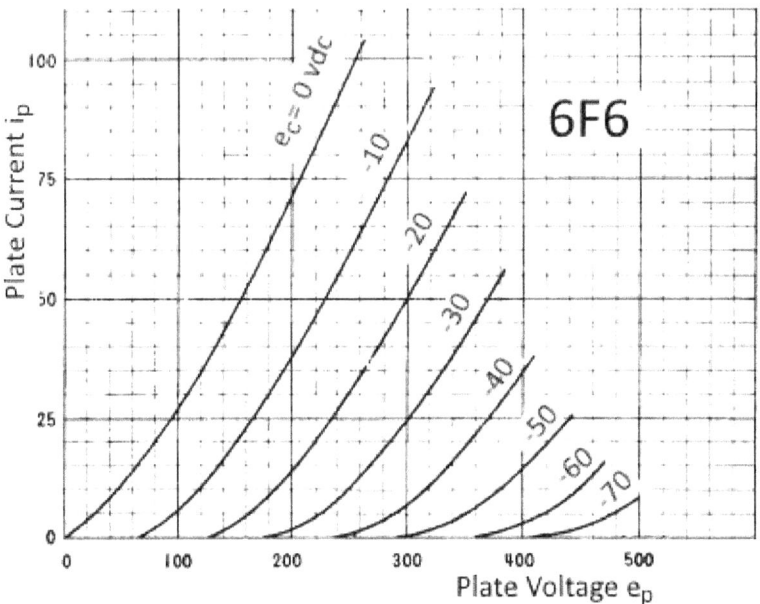

The downside of a triode-connected pentode is that we reduce its maximum power output capability, often by a factor of 5 to 1 or more. As an example, the power output of the 6F6 above is 5 watts in pentode operation. When connected as a triode, the maximum power output drops to 0.85 watts.

The upside is that the pentode's undesirable odd-order harmonic distortion disappears almost completely in triode-connected mode. We are, in effect, trading power output for less objectional harmonic distortion.

The popular 6L6 pentode is a *beam power tube* with the suppressor grid internally tied to the cathode. Instead of a screen grid, the 6L6 has beam-focusing plates that serve a similar purpose. Connecting these to the plate has an identical effect as the screen grid connected to the plate. See the plate characteristics for a triode connected 6L6 below.

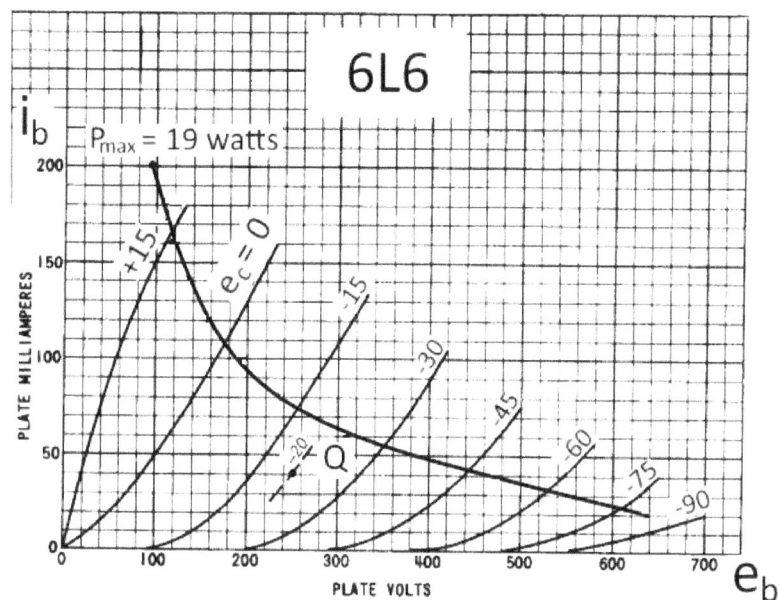

The RCA Tube Manual suggests a DC operating point Q at E_Q = 250 volts and I_Q = 40 ma. The Q point location is where the e_c = -20-volt plate characteristic would be. Using the short line shown, we calculate that r_p at that point is about 1650 Ω.

Suppose we are looking again for a distortion of 2% or less, which is an AQ to QB ratio of 1.1. Applying the steps in Chapter 4, we start with R_{ac} = $2r_p$ or 3300 Ω. See load line 1 below.

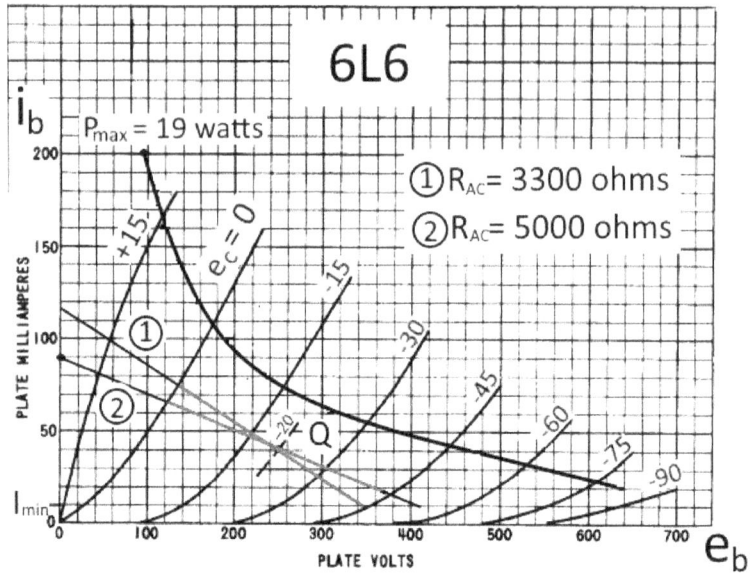

For line 1, the AQ to QB ratio is 96/78 = 1.2, which, from the distortion chart, is 4% distortion. The power output is 1.7 watts. We then try R_{ac} = 5000 Ω and draw line 2. The AQ to QB ratio is 74/67 = 1.1, giving an estimated distortion of 2%, reaching our goal. Power output has dropped only a small amount to 1.6 watts.

We need to consider two cautions when designing with triode-connected pentodes. First, check that we do not exceed the maximum screen grid voltage when connected to the plate. With some power pentodes, the maximum plate voltage is greater than the maximum screen voltage, so the limiting value in triode-connected mode is the maximum screen voltage.

For example, the 6L6's maximum plate voltage is 360 volts, but the screen maximum is 270 volts. In a triode-connected operation, the maximum plate and screen voltage is lower, at 270 volts. Exceeding this is likely to result in damage to the screen grid and tube failure. (The more recent version of the 6L6, the 6L6-GC, removes this restriction.)

Second, some triode-connected pentodes exhibit what is called *parasitic oscillation*. Power pentodes with high transconductance (g_m) can break into high-frequency oscillation (usually just above the audio range), disrupting the amplifier's normal operation.

The simple preventative is to include a *screen stopper resistor* of 100 Ω between the plate and screen and a *control grid stopper* resistor of 1 KΩ to 10 KΩ between the control grid resistor and the control grid. These modifications do not affect the tube's operation and usually eliminate parasitic oscillation.

The variety of power pentodes is quite large. An internet search reveals dozens of different pentodes used in SET amplifier design. Besides tubes like the 6L6 designed for audio applications, experimenters have used television horizontal sweep output tubes like the 6DQ6. As noted in the preface, the author's favorite is the 1625 industrial power pentode, often used in radio transmitters.

The wide variety and availability of power pentodes is an experimenter's paradise. For instance, the 300B is the Holy Grail of power triodes and can cost hundreds of dollars in "new-old stock" (NOS). For $10, we can buy a 6CM5 power pentode that, in triode-connected mode, has plate characteristics that are amazingly close to a 300B!

In the next chapter, we look at ways to bias a power output stage to obtain a specified DC operating point.

Chapter 7 – Biasing the Power Output Stage

Both the power triode and triode-connected pentode require a plate and control grid supply voltage. See the figure below.

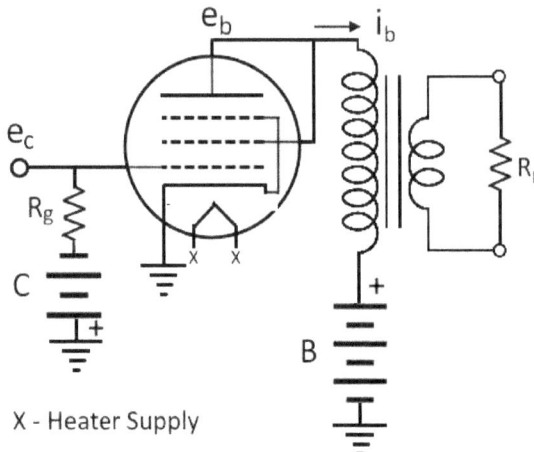

X - Heater Supply

In most designs, we derive B from a DC power supply. The B or B+ supply is usually 200 to 300 volts DC. For the control grid voltage, there are two options: *fixed-bias* and *self-bias* (sometimes referred to as *cathode-bias*).

First, we have the *fixed-bias* option shown below. It requires a negative DC supply E_c, usually in the range of -2 to -50 volts, to supply the tube's control grid bias.

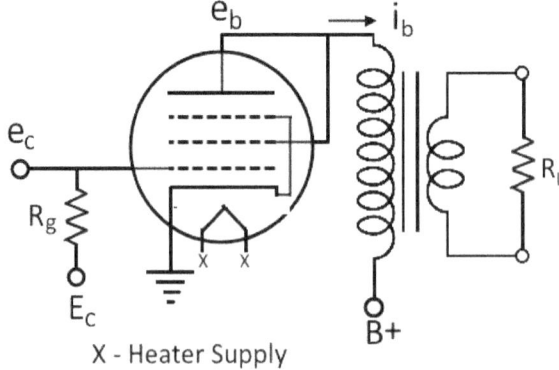

X - Heater Supply

With fixed bias, we ground the cathode and control the DC operating point using the bias voltage.

Fixed bias minimizes the components in the signal path that can reduce power output and introduce other undesirable effects, such as limiting low-frequency response. The disadvantage of fixed bias is the inconvenience of having to adjust the bias voltage both initially and as tubes age. Also, a failure in the bias supply could result in an overcurrent in the tube's plate circuit, possibly destroying the tube and output transformer!

In recent times, designers of fixed-bias power output stages have incorporated solid-state circuits that continuously monitor the plate current and adjust the bias to maintain correct DC operating conditions. We refer to this as *auto-bias*.

The second method is self-bias, for which we insert a resistor in the cathode circuit, as shown below.

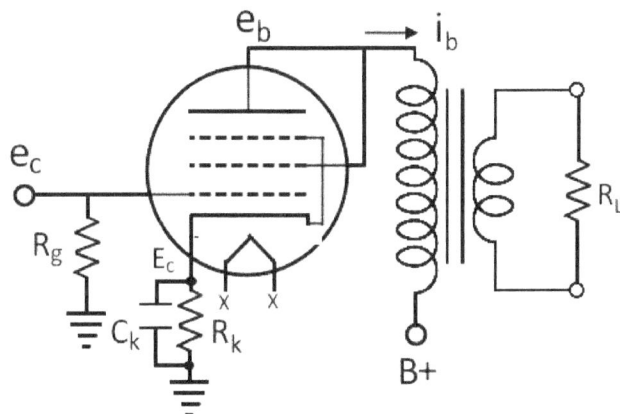

X - Heater Supply

The voltage drop across resistor E_c is positive at the cathode and, therefore, appears to be a negative voltage at the control grid. Under ordinary biasing conditions ($e_c < 0$), i_k equals i_p, so the value of resistor R_k is E_c / i_p.

Tube manuals often suggest Rk values for typical applications and DC operating points. If not, then use this approach. Assume a desired DC operating point is (I_Q, E_Q) at control grid voltage E_c. Since $i_k \approx I_Q$, R_k is simply E_c/I_Q.

In the power output stage, we cannot neglect R_k and the output transformer DC resistance R_T and must include them in the DC load line calculation. In this case, the DC load line slope is $-1/(R_T + R_k)$.

Keep in mind that these, like most calculations dealing with tubes, are based on approximate data. At best, we can expect to get in the vicinity of the DC operating point. Often, we may need to adjust values experimentally to get closer to the desired result.

To illustrate, consider the 6L6 example below.

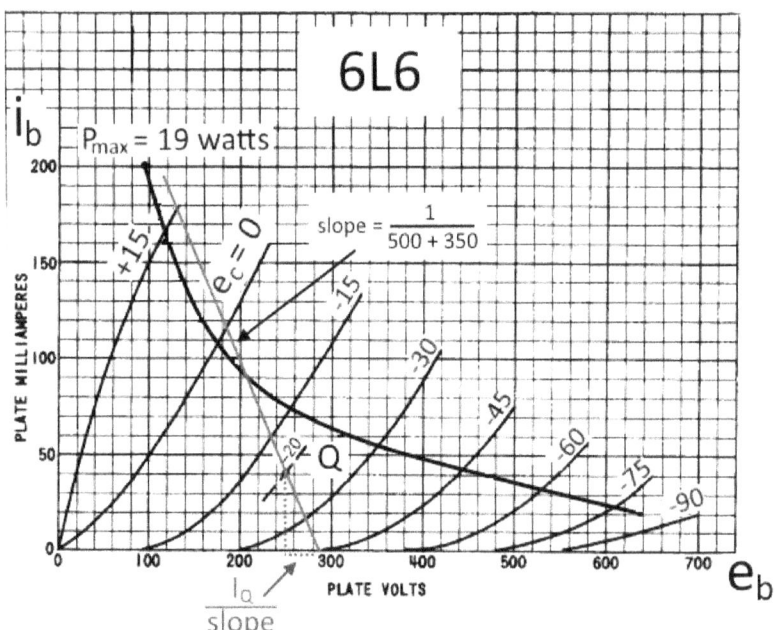

We use the DC operating point Q suggested in the tube manual. As shown, the DC load line passes through Q with a slope of $1/(R_K + R_T)$. R_K is E_c/I_Q or 10/0.04 = 500 Ω. Assume the output transformer has a DC secondary resistance R_T of 350

Ω. The B+ voltage should be E_Q+I_Q/slope or $250 + 0.04(500 + 350) = 284$ volts. The target voltage for the DC power supply is then 284 volts. Frankly, 284±5 volts makes a trivial difference overall!

The advantage of self-bias is that the presence of R_K in the cathode has a regulating effect on plate current as the tube ages. As tube emission decreases with time, so does plate current and the voltage across R_K, which increases plate current and tends to maintain the DC operating point close to its original value. Thus, with self-bias, we stabilize the DC operating point over the life of the tube.

Self-bias does have a downside. The voltage drops across R_k, and R_T reduces the plate voltage by $I_Q \cdot (R_k+R_T)$. The combined voltage drop is generally large enough to affect the DC operating point. A straightforward way to mitigate this is to raise the plate supply voltage B+ by the same amount.

Also, the cathode current flowing in R_K is equal to both the DC operating current and the audio signal current. The latter produces an audio signal voltage drop across R_K that is opposite in polarity to the plate voltage, introducing negative feedback and reducing voltage gain at audio frequencies. While neither is particularly detrimental, we specified no negative feedback in our SET amplifier. To eliminate this negative feedback, we place a suitably sized capacitor across R_K that acts as a short circuit at audio signal frequencies.

We must say that a SET purist would point out that using a self-bias resistor and bypass capacitor introduces additional components and complexity to the circuit. The simpler the design, the better it is, which is why some designers prefer fixed bias!

For the first SET amplifier, self-bias is a better choice. There are fewer things to go wrong and less maintenance overtime.

We are closing in on the design steps for our SET amplifier. We only need to tidy up a few more details. In the next chapter, we investigate design factors that affect an amplifier's overall frequency response.

Chapter 8 – SET Amplifier Frequency Response

Various components in the SET amplifier affect frequency response. As we find in this chapter, the elephant in the room is the audio output transformer. The money spent here offers the best return on investment in extending frequency response. Cost-wise, next in line is the power supply. The more robust it is, the better the low-frequency response. After these, the amplifier contains much less expensive components, including the coupling capacitor between the voltage amplifier and the cathode bypass capacitors on the voltage amplifier and power output stages.

Coupling Capacitor

To begin, let's look at the effects of the coupling capacitor between the voltage amplifier and power output stages. So far, we have assumed that we use a capacitor to couple the voltage amplifier stage output to the power output stage input. Designs exist that directly couple these stages together, but they are fraught with design and operational complications that far outweigh any benefit. So, our choice is capacitor coupled. Below is the basic triode voltage amplifier showing coupling capacitor C_c.

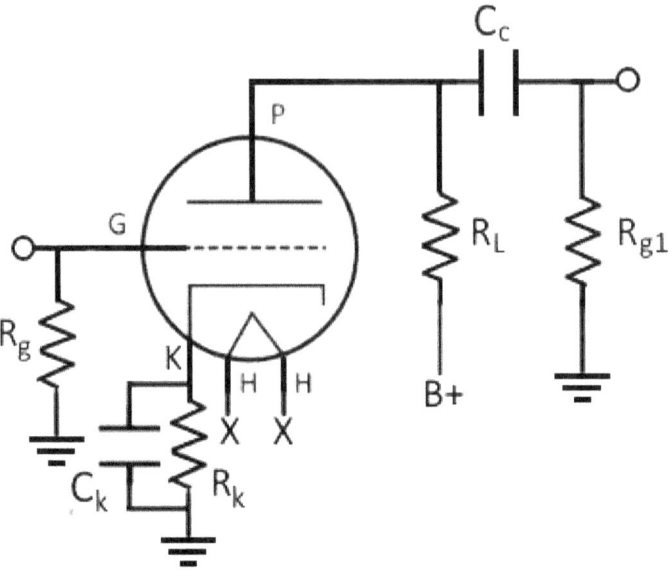

We have previously assumed that C_c is a short circuit at the operating frequencies of the amplifier. Obviously, a DC voltage change cannot pass through the capacitor, so at 0 Hz, it is an open circuit. Therefore, between the lowest operating frequency and 0 Hz, there must be a roll-off in frequency response.

To study this effect, we replace the triode above with its small signal equivalent circuit. (Yes, we should not use the small signal equivalent circuit for the SET voltage amplifier, but the analysis gives us a result that proves useful.)

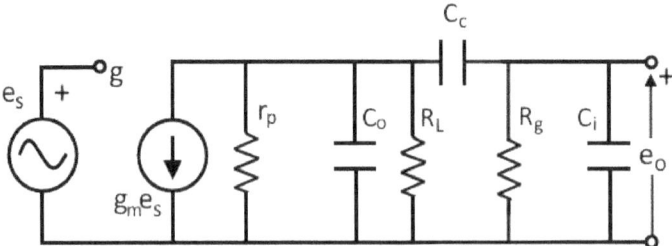

We assume that the plate supply B+ is a short circuit at signal frequencies, so we show R_L connected to the common point. The basic triode model uses a voltage-controlled ideal current source parallel to plate resistance r_p. The current source output is $g_m e_s$, where gm is the tube's transconductance and es is the control grid's audio signal voltage.

C_o is a combination of the triode internal plate capacitance plus stray wiring capacitance. R_g is the grid-leak resistor at the input to the next stage. C_i is the combined control grid and wiring capacitances of the next stage. C_o and C_i are generally small, less than 30 pF. Coupling capacitor C_c typically ranges from 0.01 µF to 0.5 µF.

We first look at mid-band gain where no frequency dependence exists. We assume that over this frequency range, C_o and C_i are open circuits, and C_c is a short circuit. The figure below shows the reduced equivalent circuit.

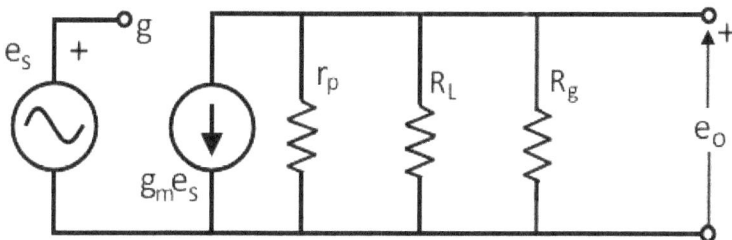

We replace the parallel combination of r_p, R_L, and R_g with equivalent value R_s where

$$\frac{1}{R_s} = \frac{1}{r_p} + \frac{1}{R_L} + \frac{1}{R_g} = \frac{r_p R_L + r_p R_g + R_L R_g}{r_p R_L R_g}$$

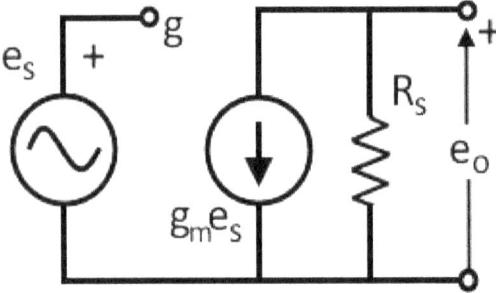

The mid-band voltage gain is then.

$$A_{mid} = \frac{e_o}{e_s} = -g_m R_s$$

The negative sign indicates that the stage inverts the input signal.

For the high-frequency response above the mid-band, we assume that C_c is a short circuit while C_o and C_i combine to produce a roll-off effect in frequency response. The equivalent circuit looks like this.

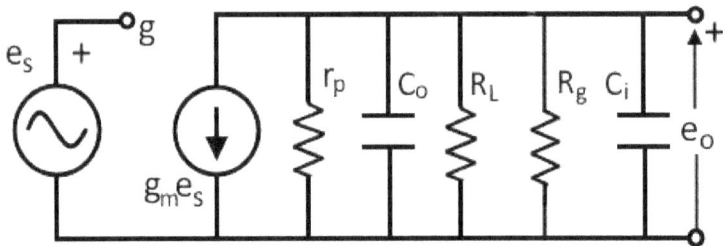

We reduce the circuit further by combining the parallel combinations of resistors and capacitors.

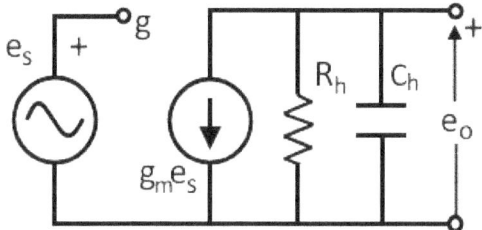

R_h equals R_s, the same as above and $C_h = C_o + C_i$. We can show that the ratio of A_{high} to A_{mid} is

$$\frac{A_{high}}{A_{mid}} = \frac{1}{1+j\omega R_h C_h}$$

We then define

$$\omega_h = \frac{1}{R_h C_h}$$

and

$$f_h = \frac{1}{2\pi R_h C_h}$$

The normalized magnitude (the ratio of A_{high} to A_{mid}) and phase angle (θ) are then given by

$$\frac{A_{high}}{A_{mid}} = \frac{1}{\sqrt{1+(\frac{f_h}{f})^2}} \qquad \theta = \tan^{-1}\left(\frac{f_h}{f}\right)$$

The figures below show the magnitude (A_{high}/A_{mid}) and phase angle (θ) plots of the high-frequency response.

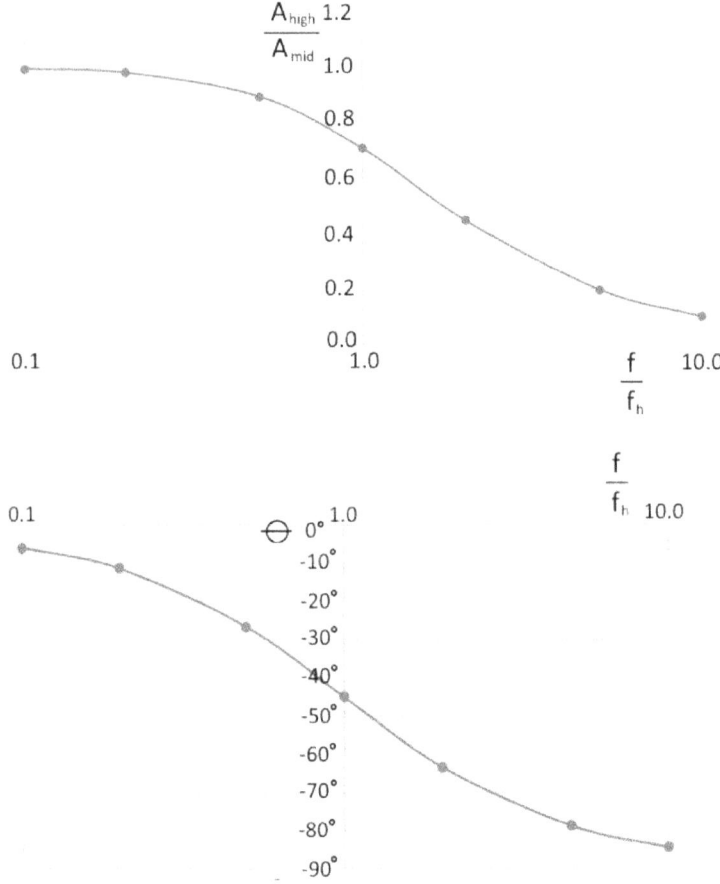

As we expected, the gain rolls off above the high-frequency value f_h. More precisely, the gain is down by a factor of 0.707 at frequency f_h. We refer to this as the *corner or breakpoint frequency*. We use "breakpoint" to refer to the frequency where the magnitude is down by a factor of 0.707 or 70.7%.

For the low-frequency equivalent circuit, we consider both C_o and C_i open circuits with only C_c affecting the frequency response. Thus, the equivalent circuit becomes

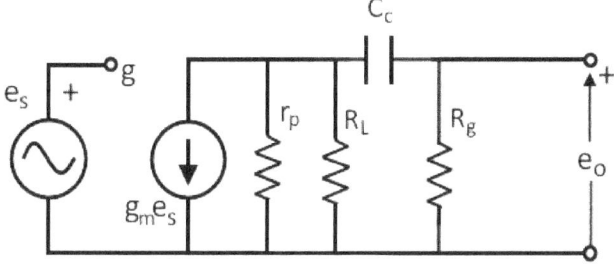

Reducing it one step further, we obtain the parallel combination of r_p and R_L.

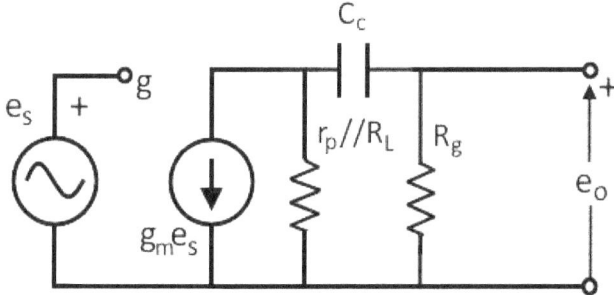

where $r_p // R_L$ is $(r_p R_L)/(r_p + R_L)$. Let

$$R_l = R_g + \frac{r_p R_L}{r_p + R_L}$$

Then, we can write

$$\frac{A_{low}}{A_{mid}} = \frac{1}{1 - j(\frac{1}{\omega C_c R_l})}$$

where

$$\omega_l = \frac{1}{C_c R_l}$$

and

$$f_l = \frac{1}{2\pi C_c R_l}$$

The figures below show the normalized magnitude and phase angle (Θ) plots of the low-frequency response.

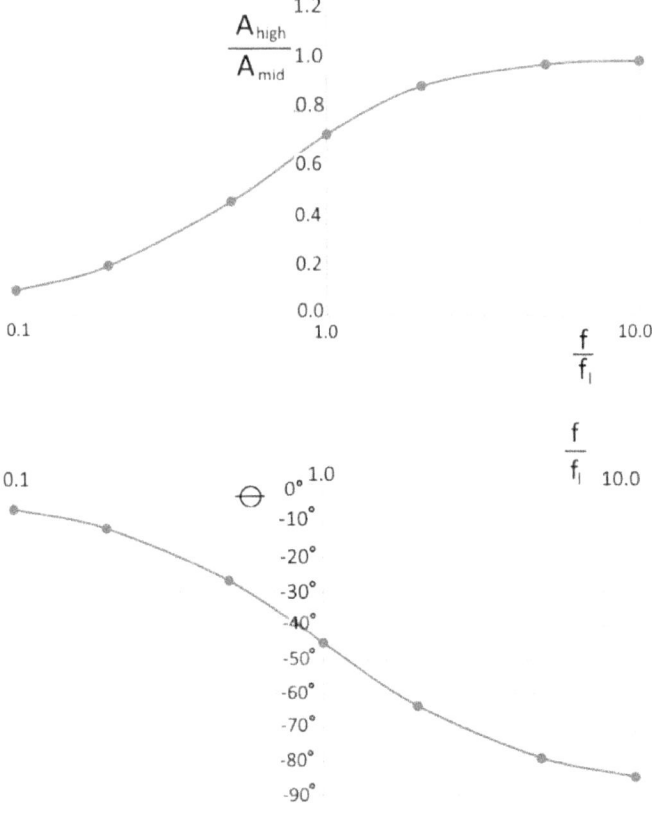

Gain magnitude is down at the breakpoint frequency f_l and falls off below that.

Combining the high and low gain magnitude plots, we see the full frequency response.

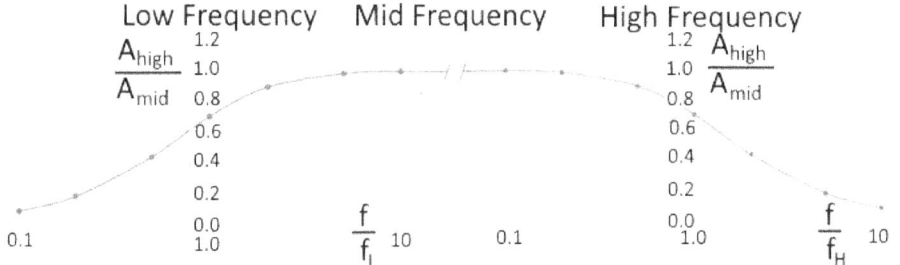

Note that at f equal f_l and f equal f_h, the gain is down 0.707, which are the mib-band breakpoint frequencies. Frequencies f_l and f_h establish limits of the amplifier's low to the high-frequency range. Based on this, we want f_l to be much less than the specified frequency response at the lower end of the stage. Likewise, f_h should be very much greater than the upper end.

As a typical example, suppose a voltage amplifier has these parameters

r_p = 80,000 Ω
R_L = 220,000 Ω
R_g = 470,000 Ω
C_o = C_i = 30 pF (tube plus wiring)

We want to find a suitable value for coupling capacitor C_c. Assume the frequency response of our amplifier is 20 to 20,000 Hz. Then, we want f_h>>20000 Hz and f_l << 20 Hz. Let's check the high-frequency end first.

R_h = 80,000//220,000//470,000 = 52,156 Ω
C_h = 30x10^{-12} F

$$\frac{1}{2\pi \cdot 52{,}156 \cdot 30 \cdot 10^{-12}} = 101{,}000 \, Hz \gg 20000$$

So, we are in decent shape on the high-frequency end. Now, let's see what size coupling capacitor C_c we should use.

R_l = 470,000 + 80,000//220,000 = 470,000 + 58,670 = 528,670

We want

$$f_l = \frac{1}{2\pi \cdot 528{,}670 \cdot C_c} \ll 20$$

Solving the inequality, we have

$$C_c \gg \frac{1}{2\pi \cdot 528670 \cdot 20} = 0.015 \, \mu F$$

Choosing C_c ten times the minimum C_c = 10 x 0.015 = 0.15 ≈ 0.1 µF should work fine.

Cathode Bypass Capacitor

Next, we consider the effect of cathode bypass capacitor C_k on frequency response in the voltage amplifier circuit below.

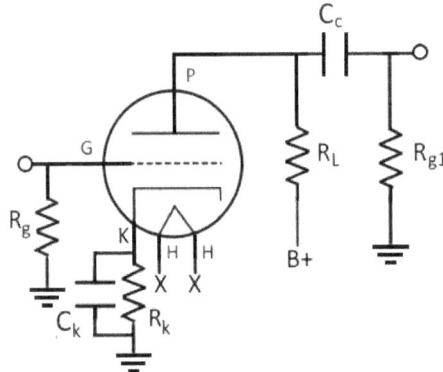

At mid-band, C_k is a short circuit, and the amplifier's gain is maximum. As we lower the frequency e_s, the impedance of C_k begins to decrease, becoming essentially an open circuit at 0 Hz (DC). At exceptionally low frequencies, R_k is essentially un-bypassed, and the amplifier's gain reaches a minimum. Our task is to make certain that this range of changing gain is much less than the amplifier's low-frequency specification.

To attack this problem, we use the equivalent circuit below.

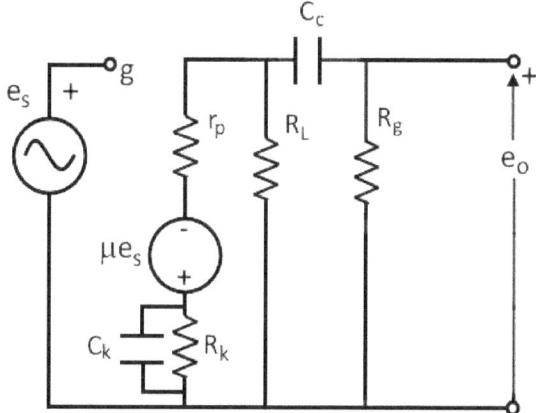

The derivation of gain is way too complex, so we cut to the chase and show the result.

$$\left[\frac{A_{low-k}}{A_{mid}}\right]_{min} = \frac{r_p + R_L}{r_p + R_L + R_k(1+\mu)}$$

This expression is the minimum gain when the frequency is so low that the cathode bypass capacitor acts as an open circuit. At high frequencies, the bypass capacitor acts as a short circuit and the R_k term disappears, leaving the normalized mid-band gain as 1.

Let's now investigate the two frequency limits that mark the transition between the cathode bypass capacitor's minimum and mib-band gain limits. First, make the following definitions.

$$f_{kl} = \frac{1}{2\pi C_k R_k}$$

and

$$f_{kh} = \frac{1 + \frac{R_k(1+\mu)}{(r_p + R_L)}}{2\pi C_k R_k}$$

Note that the denominators are the same, and the numerator of f_{kh} is larger than that of f_{kl}, so $f_{kh} > f_{kl}$. The general expression for the normalized gain, then, is

$$\frac{A_k}{A_{kmid}} = \frac{f_{kl}}{f_{kh}} \cdot \frac{1 + j\frac{f}{f_{kl}}}{1 + j\frac{f}{f_{kh}}}$$

The normalized magnitude and phase (Θ) are

$$\frac{A_k}{A_{kmid}} = \frac{f_{kl}}{f_{kh}} \cdot \sqrt{\frac{1 + \left(\frac{f}{f_{kl}}\right)^2}{1 + \left(\frac{f}{f_{kh}}\right)^2}} \qquad \theta = \tan^{-1}(\frac{f}{f_{kl}}) - \tan^{-1}(\frac{f}{f_{kh}})$$

Plotting the magnitude and phase for the low-frequency response for a typical amplifier, we have

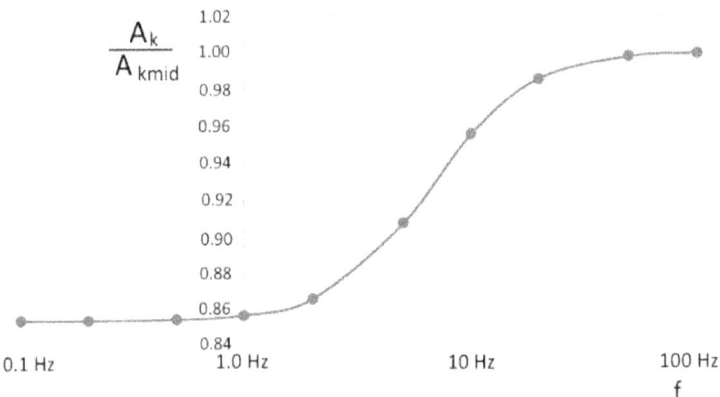

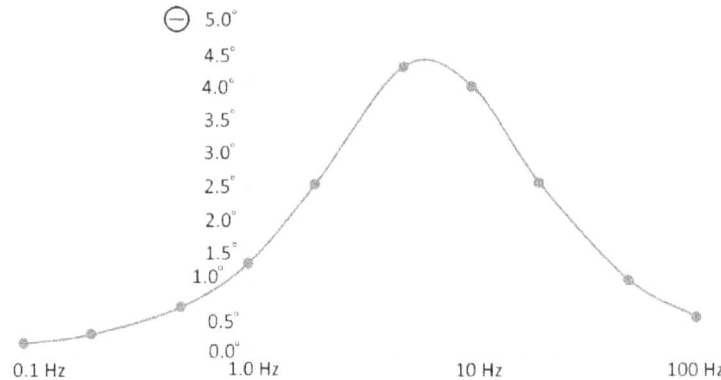

Note that as the frequency drops below ~100 Hz, (1) the normalized magnitude falls off from 1 and levels off just below 0.86, and (2) the phase shift rises only slightly to ~4.5° and then falls back to zero. Overall, the effect is minimal, so the only real consideration is to make f_{kh} << the low-frequency specification.

As an example, suppose R_k = 2200 Ω, and the low-frequency specification is 20 Hz. We choose f_{kh} one-tenth of 20 Hz or 2 Hz. Solving the equation for f_{hk} for C_k, we get

$$C_k = \frac{1 + \frac{R_k(1+\mu)}{(r_p + R_L)}}{2\pi f R_k}$$

Using r_p = 80000 Ω, μ = 100, and R_L = 220 KΩ, we find that C_k = 63 µF. Probably, a standard electrolytic value of 47 µF would do.

For power triodes, we cannot use the equivalent circuit above. We want the cathode bypass capacitor C_k to act as a short circuit at the amplifier's low-frequency specification F_{low}. We can readily achieve this by choosing C_k so that the following is true.

$$\frac{1}{2\pi R_k C_k} = f_k \ll F_{low}$$

Given F_{low} and solving for C_k, we have

$$C_k = \frac{1}{2\pi R_k f_k}$$

As an example, if R_k = 270 Ω and f_k = 20/10 = 2 Hz, then C_k would be 294 µF. A standard value of 250 µF would work.

The third component that affects frequency response is the audio output transformer. Its purpose, as already mentioned, is to match the exceptionally low impedance of the speaker (4 to 16 Ω) to a high impedance related to the choice of power output tube and the amplifier specification (power output and distortion).

Output Transformer

An ideal transformer couples equally well across all frequencies. A real transformer falls short, typically exhibiting a frequency response not unlike that observed above; that is, a relatively flat mid-band falling off at frequencies above and below.

To study these effects, we introduce an equivalent circuit for the transformer.

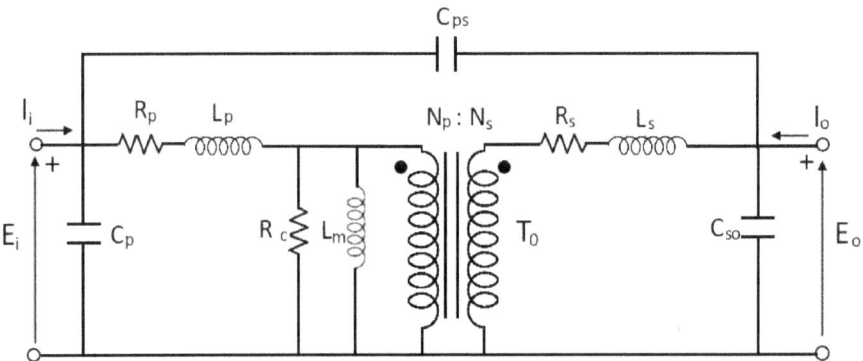

With these components,

T_0 - An ideal transformer with primary-to-secondary turns ratio n = N_p/N_s with no energy loss. The placement of the two dots indicates polarity at that point. In the figure, the dot's polarity is the same at the upper end of the transformer; i.e., no inversion occurs between primary and secondary.

C_p & C_{so} - The lumped* capacitance between successive turns of wire in the primary and secondary windings, respectively.

R_p & R_s - The DC resistance of the primary and secondary, respectively.

L_p & L_s - The primary and secondary coils link with each other to produce the transformer effect. However, the linkage is not 100%. The lost linkage produces an undesirable voltage drop, acting like an inductor in series with the input and output. This leakage inductance is captured by introducing inductors L_p and L_s.

L_m - The shunt magnetizing inductance of the primary.

R_c - This resistor introduces a loss equivalent to eddy-current and hysteresis losses in the transformer's core of ferromagnetic material.

C_{ps} - The capacitive coupling between the primary and secondary windings.

* The term *lumped* refers to the fact that while these capacitances distribute across the windings, we use a single capacitance to approximate their effect. We should note that this and other approximations make the equivalent circuit useful, though it is not an exact working model!

The presence of C_{ps} complicates the analysis, so the first step is to move its effect into the capacitor C_{so} by replacing C_{so} with capacitor C_s, where

$$C_s = C_{so} + (1 - \frac{N_p}{N_s}) \cdot C_{ps}$$

The reduced equivalent circuit becomes

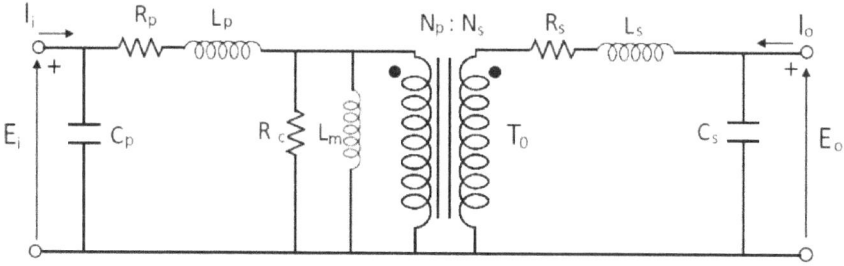

The next step is to refer the secondary components to the primary side, which gives this result.

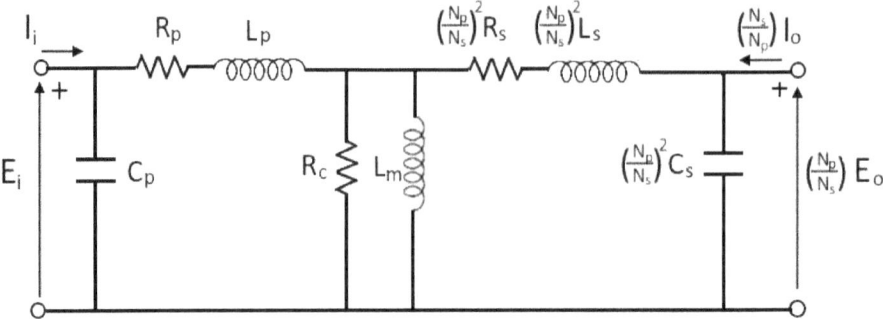

With this, we are ready to study how the audio output transformer affects frequency response.

To study transformer frequency response, we focus our attention on current gain. Recall that the transformer's function is to take a relatively small signal current at the plate of the power triode and transform it to the large current needed to drive the low-impedance speaker. Ideally, the output current to the speaker $i_{out} = \left(\frac{N_p}{N_s}\right) \cdot i_{in}$.

From this, we see that the ideal current gain in the transformer is $A_i = (N_p/N_s)$ or simply the ratio of primary to secondary turns. For a real transformer, this gain varies over the audio frequency range, falling off at both ends of the audio range and sometimes presenting irregularities in the mid-range region.

To begin the discussion, let's look at a professionally designed transformer's mib-band region where we assume C_p, C_s, L_p, and L_s are short circuits, and L_m is an open circuit. We have also eliminated R_c, as it is generally large enough to ignore in most cases.

Given that the plate resistance of the power triode is r_p and the load impedance on the transformer is R_L, the equivalent circuit above reduces to this.

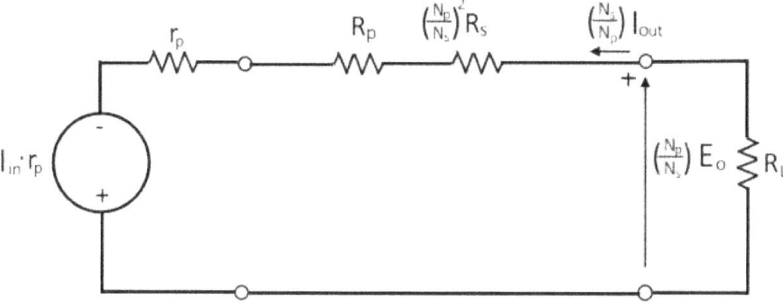

Solving a simple loop equation, we find that the mid-band current gain is

$$A_{imid} = \frac{N_p}{N_s} \cdot \frac{r_p}{r_p + R_p + (\frac{N_p}{N_s})^2 \cdot (R_s + R_L)}$$

This equation suggests that by keeping winding resistances R_p and R_s small, we maximize mid-band gain. R_s is not a problem, as the secondary has very few turns compared to the primary. Practical limits related to primary coil size keep its wire gauge small while the number of turns is large. Nevertheless, high-quality transformers keep R_p less than 100 Ω so that its effect on mib-band response is not significant.

To explore low-frequency response, we use the equivalent circuit below.

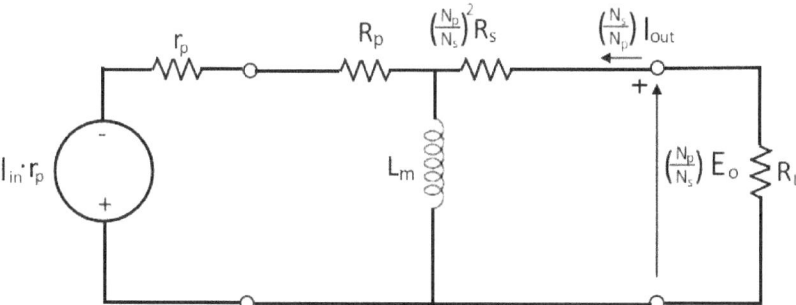

The leakage inductances L_p and L_s are small and essentially short circuits at low frequencies, so that we can neglect them in the circuit. The transformer core inductance L_m is significant as it short circuits the signal path, as shown. Circuit analysis yields this result for the magnitude of the circuit's low-frequency current gain.

$$\frac{A_{ilow}}{A_{imid}} = \frac{1}{1 - j(\frac{f_l}{f})}$$

where $f_l = R_{low}/2\pi L_m$ and

$$R_{low} = \frac{(r_p + R_p)[(\frac{N_p}{N_s})^2 \cdot (R_s + R_L)]}{r_p + R_p + [(\frac{N_p}{N_s})^2 \cdot (R_s + R_L)]}$$

Note that this is the same low-frequency response we saw with the capacitor-coupled amplifier, where f_l is the low-frequency breakpoint. As we reasoned before, we want f_l to be much less than the amplifier's low-frequency specification. We can do little to reduce R_{low} as its component values are not readily changeable. Increasing the value of L_m is our only choice.

Impedance L_m is related to the mass of ferromagnetic material in the core. More "iron" means a heavier, bulkier, and more costly transformer. A low-fidelity transformer in a consumer radio may only have an L_m of 2 H (Henries), while an expensive HiFi transformer could sport an L_m of 30 H or more!

An example illustrates the low-frequency response of a quality audio output transformer. A transformer for the 2A3 power triode has these characteristics: $R_p = 70\ \Omega$, $R_s = 1\ \Omega$, $L_m = 20$ H, and $(N_p/N_s)^2 = 300$. Assuming the r_p for the 2A3 is 800 Ω, then plugging into the formula for flow, we get 5.2 Hz, which is a reasonable value given a frequency response low specification of 20 Hz.

Next, we use the circuit below to study a transformer's high-frequency response.

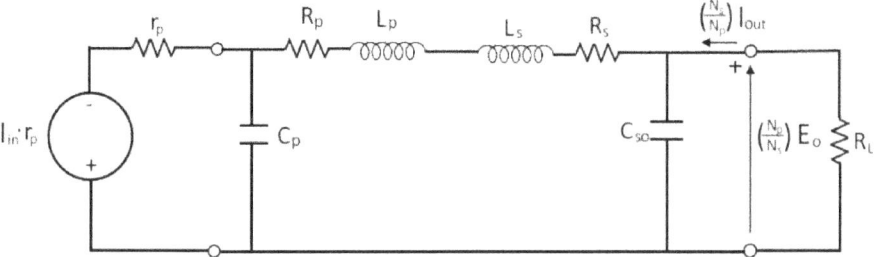

The large impedance of the core inductance L_m at high frequencies disappears, and the two leakage inductances, L_p and L_s, become the limiting factor because they are in series with the signal current. A reasonable assumption over the audio frequency range is that the following holds.

$$\left(\frac{N_p}{N_s}\right)^2 R_L \ll \frac{1}{2\pi C_p}$$

and

$$R_L \ll \frac{1}{2\pi C_{pso}}$$

The analysis then shows the high-frequency current gain to be

$$\frac{A_{ihigh}}{A_{imid}} = \frac{1}{1 + j(\frac{f}{f_h})}$$

where $f_h = R_T / (2\pi L_T)$ and

$$R_T = r_p + R_p + [\left(\frac{N_p}{N_s}\right)^2 \cdot (R_s + R_L)]$$

$$L_T = L_p + \left(\frac{N_p}{N_s}\right)^2 \cdot L_s$$

Again, we can do little to reduce R_T, so the focus is on L_p and L_s. Careful manufacturing techniques can reduce these so that f_h is much greater than the high-frequency specification of the amplifier.

Let's look at an example. Suppose the measured value for $L_p + L_s$ for a low-end audio output transformer is 52 mH. The remaining parameters are as follows: R_p = 70 Ω, R_s = 1 Ω, R_L = 8 Ω, and $(N_p/N_s)^2$ = 625. From the formula above, we find that the breakpoint is 22,000 Hz, which is not suitable for HiFi since unacceptable attenuation occurs in the 15,000 to 20,000 Hz range.

A similar but high-quality audio transformer has a lower leakage inductance of 11 mH. The resulting breakpoint rises to 104 KHz with minimal attenuation at 20,000 Hz. Out of interest, the transformer of the paragraph above costs $16, while this one costs $50.

Lastly, we must consider the effects of DC operating current and power handling capability. With too much current, the transformer's magnetic flux reaches a maximum and transformer action stops with additional audio signal current. In our SET amplifier, the DC operating current should raise the flux to roughly the center of its range, whereby the signal current for the delivered power can rise and lower flux within its functioning range.

In summary, a quality audio output transformer has the following characteristics:

1. Low primary and secondary DC resistance, usually less than 100 and 1 Ω, respectively.
2. Low leakage inductance, usually less than 10 mH total primary and secondary.
3. High core inductance, usually 10 H or more.
4. Adequate DC operating current and power handling capability.

We do not always have access to this information. Still, an audio output transformer with a suitable power rating, matching impedance, and DC operating current that specifies a frequency response ±1 db over 20 to 20,000 Hz should do the job.

Lastly, let's consider the effect of the power supply on the frequency response of the SET amplifier. An ideal power supply would provide a constant voltage as the frequencies of the audio signal load current change. At mid and high audio frequencies, this is not generally a problem. At low frequencies below 100 Hz, the power supply might struggle to supply current during the longer periods of elevated current. Low frequencies can cause the power supply average voltage to drop, which, in turn, reduces stage gains and introduces what can be severe distortion. In a worst-case scenario, the audio signal flattens and becomes decidedly misshapen. Therefore, the design of the power supply must have sufficient current supply capacity so that it approximates an ideal voltage source in the low frequencies specified.

That about covers all the design elements of our SET amplifier. In the next chapter, we tackle a complete SET design example.

Chapter 9 – Example SET Amplifier Design

Having established the SET amplifier theory, we are able to carry out a design. The circuit below combines the voltage amplifier and power output stages described in the previous chapters.

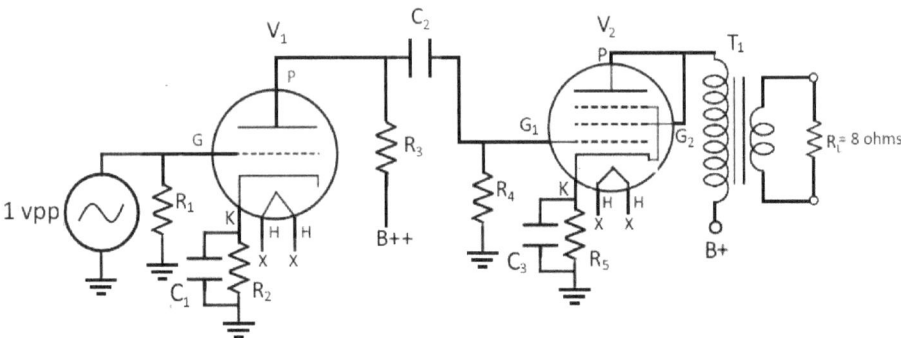

The specifications for the design are as follows: (1) A 0.5 vpp 1000 Hz input delivers 1-watt output into an 8-Ω speaker. (2) Frequency response is 30 to 15,000 Hz ±1 db. (3) Maximum harmonic distortion is 4% at 1000Hz. (5) Input load is 100 KΩ minimum.

Based on these specifications and the theory developed thus far, we can determine the component values in the circuit. We cannot determine B+ and B++ voltages until after we design the power output stage.

Note that I have chosen to use a triode-connected pentode rather than a triode. Our basis for this is (1) The triode-connected pentode configuration is simpler to implement than a filament power triode, and (2) A wide selection of power pentodes are readily available and economically priced.

Step 1 – Choose a power output tube.

When choosing a power pentode, the 6L6 would be a logical choice, as we have studied it closely in previous chapters. Rather than that, let's pick a different tube and carry out a complete design starting from scratch.

As mentioned in the Preface, my first hi-fi amplifier was a Knight Kit 12-watt amplifier circa 1960. It boasted push-pull 6BQ5 power pentodes. Amazingly, 60 years later, manufacturers still produce the 6BQ5 (EL84, its European designation)! A little nostalgia on my part, coupled with the 6BQ5's low cost and availability, made it my personal choice for our SET amplifier design.

The first thing to do is check the 6BQ5's pentode power output. It should be at least five times our specification of 1 watt. Consulting a tube manual, the maximum power output for the 6BQ5 is 5.7 watts. So, we are good here.

The next thing is to find 6BQ5 triode plate characteristics. We won't be able to carry out the design procedure without them. A check of tube manuals at Frank's electron Tube Data sheets (https://frank.pocnet.net) yielded triode plate characteristics and recommended circuit values in the Phillips Tube Manual (>1964). See the figure below.

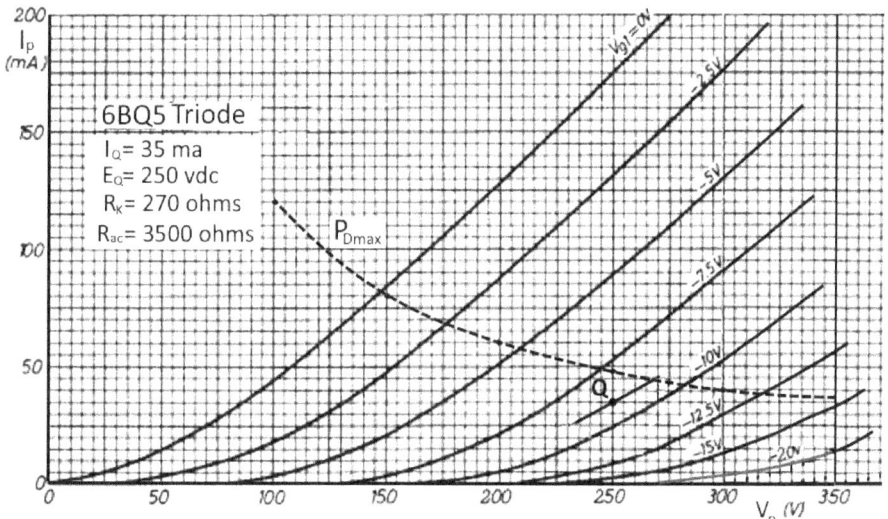

The first thing to notice is the irregular spacing of the plate curves about the recommended DC operating point Q. The 6BQ5's plate curve spacing varies much more than the 2A3 and triode-connected 6L6, which means greater harmonic distortion and possible difficulty in achieving the 4% target. Nevertheless, let's proceed and see what happens.

[Parts List Choice: V2 JJ Electronics 6BQ5 AES* T-EL84-JJ**]

* AES is Antique Electronic Supply (tubesandmore.com).
** T-EL84-JJ is the part number.

Step 2 – Choose an Output Transformer.

To choose an output transformer, start by calculating plate resistance r_p at the recommended DC operating point.

On the plate characteristics, draw a right triangle using a short line C-D drawn parallel to the adjoining plate curves at the recommended DC operating point. (Note: When working with plate characteristics, it is best to print them enlarged to the size of a standard sheet of paper. The figures shown here are too small to give accurate results. They are for illustration purposes only.)

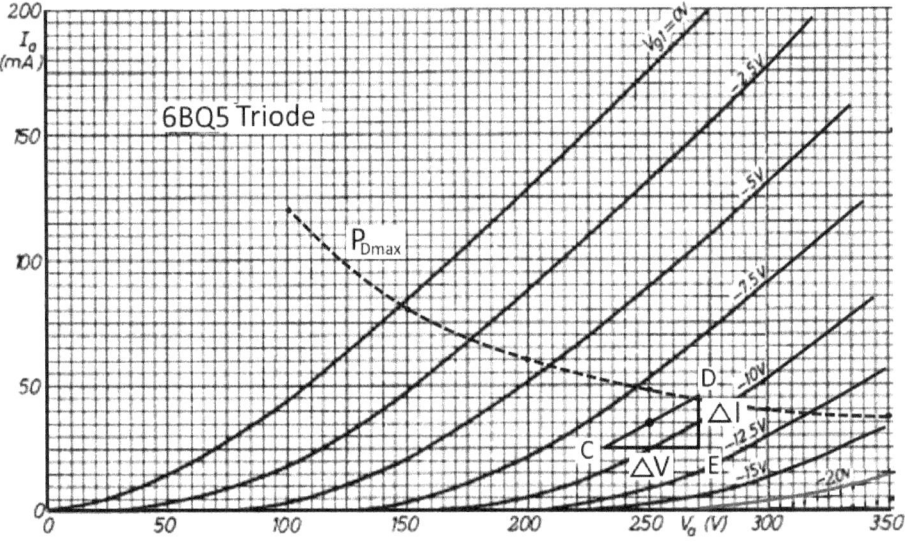

From the sides of the triangle, determine ΔV and ΔI and compute $r_p = \Delta V / \Delta I$ = 33 vdc /.020 A = 1650 Ω.

Next, plot trial AC load lines for R_{ac} = 1650 Ω, 2 · 1650 = 3700 Ω, and 3 · 1650 ≈ 5000 Ω. See the figure below.

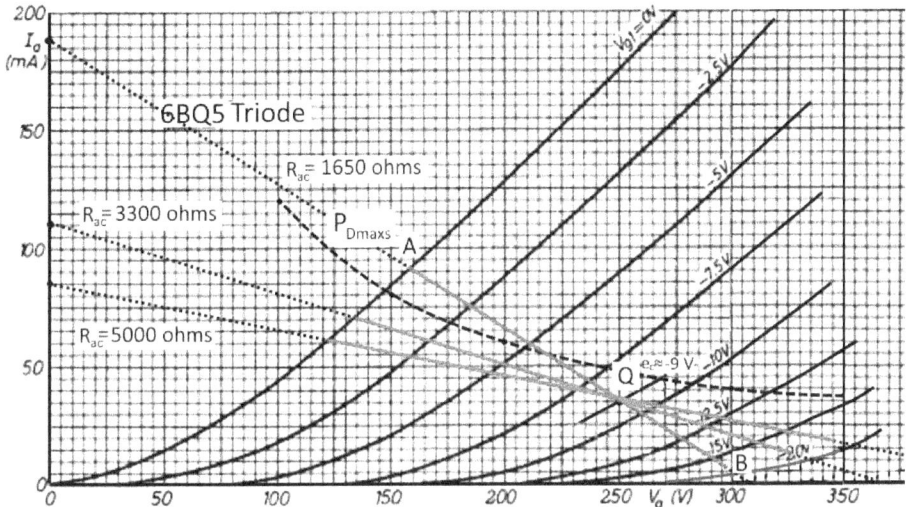

In each case, the voltage swing of e_c is twice the value of e_c at Q. We must estimate the value since it falls between steps e_c = -7.5 vdc and -10 vdc. A value of -9 vdc is a good estimate. The e_c voltage swing is then 0 vdc to -18 vdc. Here again, -18 vdc is not a step value, so stop the swing just short of e_c = -20 vdc.

With the voltage swing line segments marked, we can estimate harmonic distortion and power output. To estimate harmonic distortion, we measure line segments AQ and QB for each load line. From these values, we compute the ratio AQ / QB and read the estimated harmonic distortion from the chart below.

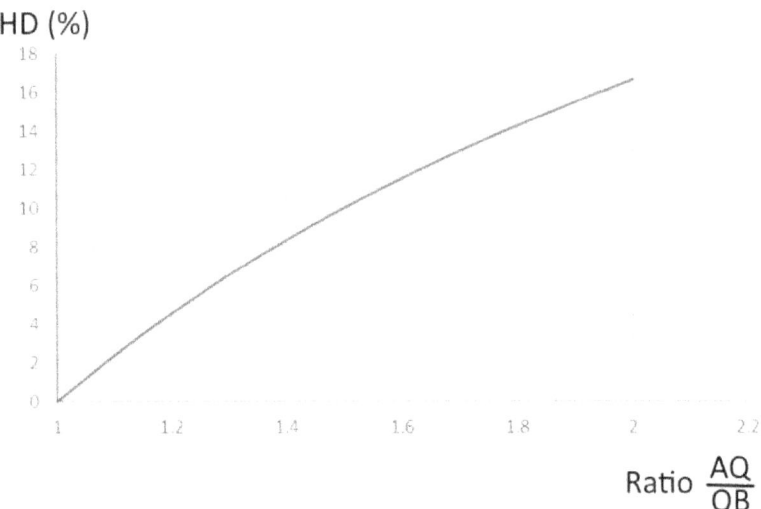

To estimate the power output, use this equation:

$$P_{out} = \frac{\Delta I_p \cdot \Delta E_p}{8} \text{ watts} = \frac{(I_B - I_A) \cdot (V_B - V_A)}{8} \text{ watts}$$

Here are the results for the three values of R_{ac}.

$R_{ac} = r_p = 1650\ \Omega$:

$R_{ac} = 1650\ \Omega$, AQ = 95 mm, QB = 48 mm, THD ≈ 16%
ΔV = 139 vpp, ΔI = 84 mapp, P_{out} ≈ 1.5 watts

We knew from theory not to expect $R_{ac} = r_p$ to be a desirable choice. The power output is good, but 16% harmonic distortion is unacceptable.

$R_{ac} = 2 \cdot r_p = 3300\ \Omega$:

$R_{ac} = 3300\ \Omega$, AQ = 106 mm, QB = 69 mm, THD ≈ 10%
ΔV = 193 vpp, ΔI = 62 mapp, P_{out} ≈ 1.5 watts

Twice r_p is a better choice, but the distortion is still too high. Again, from theory, we know that increasing R_{ac} further reduces distortion at the expense of power output. At 1.5 watts, there is room to work before getting to the specified 1 watt.

$R_{ac} = 3 \cdot r_p \approx 5000\ \Omega$.

$R_{ac} = 5000\ \Omega$, AQ = 113 mm, QB = 85 mm, THD ≈ 6%

$\Delta V = 220$ vpp, $\Delta I = 43$ mapp, $P_{out} \approx 1.2$ watts

With a small loss of power output, this choice brings the harmonic distortion down to a marginally acceptable 6%. Increasing R_{ac} further would reduce the THD, but the loss in power output would likely drop below the specified 1 watt.

At this point, it would make sense to rethink the choice of power pentode and change to one offering a higher power capability (like the 6L6-GC). Given the approximate nature of the calculations, perhaps it's too early to change. So, we keep the 6BQ5 for now and use the breadboard test in the next chapter to sort out the harmonic distortion question.

EDCOR makes a quality 5000 Ω to 8-Ω output transformer that accommodates the 35-mA plate current. It also works well with a 6L6-GC if we switch later.

[Parts List Choice: T1 5000: 8-Ω 10 W 80 mA output transformer EDCOR* GXSE10-5K]

* EDCOR is EDCOR Electronics Corporation (edcorusa.com).

Step 3 – Choose Power Output Stage Passive Components.

From the Phillips Tube Manual, the recommended self-bias resistor R_5 is 270 Ω. Note that with this resistor, e_c is -270·0.035 ≈ -9 vdc, which is consistent with the estimated value of e_c taken from the position of Q in the plate characteristics. Had it not, an adjustment in its value would have been in order.

Since $I_Q = 35$ ma, the power dissipated by R_5 is $I^2 \cdot R = (0.035)^2 \cdot 270 = 0.33$ watts. While a ½ watt resistor would do, making it 1 watt ensures that it runs cool.

[Parts List Choice: R_5 270-Ω 1-watt carbon film resistor AES R-B270]

To set the value of the cathode bypass capacitor C_3, choose a frequency f_{low} to be below the specified lower limit of 30 Hz. How much would we choose below? One-tenth is a reasonable factor. So, choosing a value one-tenth of 30 Hz, f_{low} = 30/10 = 3 Hz. Calculate C_K from this formula.

$$C_K = \frac{1}{2 \cdot \pi \cdot f_{low} \cdot R_K} = \frac{1}{2 \cdot \pi \cdot 3 \cdot 270} > 196 \; \mu F$$

A readily available 250 µF electrolytic capacitor is an excellent choice. To get its voltage requirement, note that the drop across R_5 is the grid bias voltage 9 vdc. A 250 uF at 25 vdc electrolytic capacitor works fine.

[Parts List Choice: C_3 250 uF 25 vdc electrolytic capacitor AES C-ET250-25-MOD]

The tube manual specification indicates that, for self-bias operation, the maximum value for the 6BQ5's control grid resistor is 1 MΩ. Making R_5 a 470 KΩ ½ watt seems reasonable.

[Parts List Choice: R_4 470 KΩ ½ watt carbon film resistor AES R-A470K]

Step 4 – Calculate B+ and B++ Voltages.

If E_Q is to be 250 vdc, then the B+ voltage calculation must consider the voltage drops across the self-bias resistor R_5 and the DC resistance of the output transformer's primary winding. Both resistances are in series with the plate-to-cathode circuit of the 6BQ5 and reduce the effective plate voltage. Given that R_5 is 270 Ω, and, from the EDCOR specification, the DC resistance of the audio output transformer primary winding resistance is 180 Ω, the required B+ voltage is 250 + (270 + 180) · 0.035 = 266 volts.

For now, assume that B++ = 266 vdc, also. In the power design chapter, we look more closely at requirements for both B+ and B++, which could produce a different value for B++.

Step 5 – Choose a Voltage Amplifier Tube.

The choice here of a 12AX7 is obvious. Like the 6BQ5, the widely used 12AX7 is currently manufactured, readily available, and reasonably priced.

[Parts List Choice: V1 JJ Electronics 12AX7 AES T-12AX7-S-JJ]

Step 6 – Choose Voltage Amplifier Components.

Looking back at the 6BQ5 AC load line, note that full output requires an 18 vpp signal ($2 \cdot |e_c| = 2 \cdot 9 = 18$ vpp) as the voltage amplifier output. For the specified input of 0.5 vpp, the voltage gain required is then $18 / 0.5 = 36$. Generally, it is good practice to design for a larger voltage gain and then use a potentiometer volume control to reduce the gain as desired. So, choose $A_v = 60$.

As indicated previously, most tube manuals provide suggested component values for different voltage amplifier requirements. We use the chart below from the RCA Tube Manual (1975).

B+	R_p	R_g	R_k	C_k	C_c	E_o	A_v	
180	0.1	0.1	1800	—	4.0	0.025	18	40	
	0.1	0.22	2000	—	3.5	0.013	25	47	
	0.1	0.47	2200	—	3.1	0.006	32	52	
	0.22	0.22	3000	—	2.4	0.012	24	53	3AV6
	0.22	0.47	3500	—	2.1	0.006	34	59	4AV6
	0.22	1.0	3900	—	1.8	0.003	39	63	6AV6
	0.47	0.47	5800	—	1.3	0.006	30	62	6EU7*
	0.47	1.0	6700	—	1.1	0.003	39	66	12AV6
	0.47	2.2	7400	—	1.0	0.002	45	68	12AX7A/
300	0.1	0.1	1300	—	4.6	0.027	43	45	ECC83*
	0.1	0.22	1500	—	4.0	0.013	57	52	20EZ7*
	0.1	0.47	1700	—	3.6	0.006	66	57	7025*
	0.22	0.22	2200	—	3.0	0.013	54	59	
	0.22	0.47	2800	—	2.3	0.006	69	65	
	0.22	1.0	3100	—	2.1	0.003	79	68	
	0.47	0.47	4300	—	1.6	0.006	62	69	
	0.47	1.0	5200	—	1.3	0.003	77	73	
	0.47	2.2	5900	—	1.1	0.002	92	75	

* One triode unit.   • Peak volts.

(See https://archive.org/details/RCA_RC-30_1975/page/n647/mode/2up.)

Design B+ = 266 vdc is between the entries in the chart. Choose the closest B+ option, which is B+ = 300 vdc. Within this option, choose the line with the closest A_v greater or equal to 60. Choose the line with $A_v = 65$. From the chart, this makes $R_2 = 2700\ \Omega$ and $R_3 = 220\ K\Omega$.

To verify that these values provide the required gain, plot the load line on an enlarged paper copy of the 12AX7 plate characteristics. Find these at www.github.com/rbwhipple/SET_Book. See the figure below.

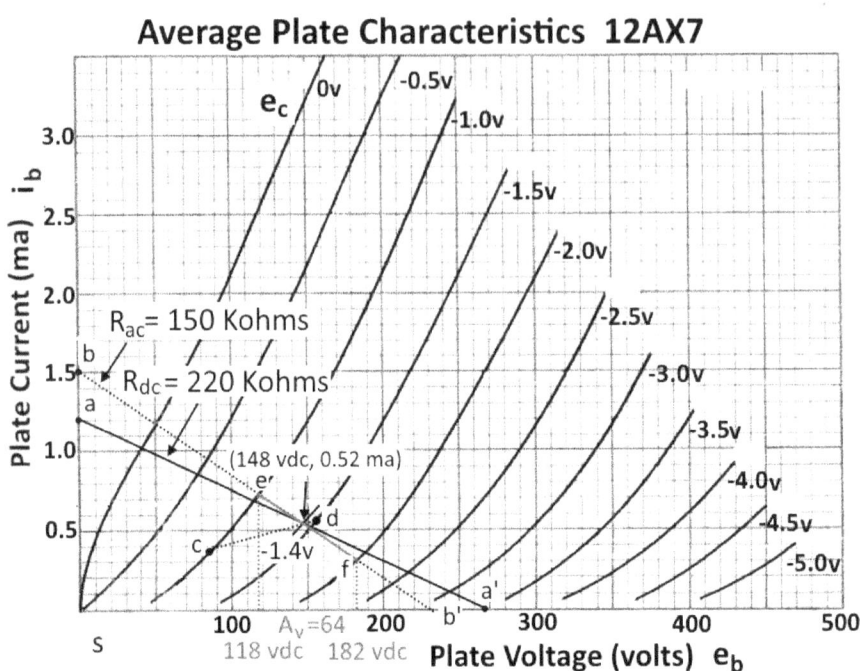

To get the DC operating point, plot e_c versus i_b for the 2700-Ω resistor. Point "c" is 1.0 / 2700 = 0.37 ma, and point "d", 1.5 /2700 = 0.56 ma. Drawing a straight line between points "c" and "d," where it crosses the load line, is the estimated DC operating point (148 vdc, 0.52 ma). The value of grid bias voltage e_c at that point is -0.52 · 2700 = -1.4 vdc.

To estimate the voltage gain, plot the AC load line. Assuming the coupling capacitor is a short circuit at 1000 Hz, R_{ac} is the parallel combination of the 220 KΩ and 470 KΩ resistors or 150 KΩ. We plot the AC load line as line b – b'.

Next, find the change in plate voltage as e_c varies about the DC operating point along the AC load line. The plate curves for e_c = -1.0 to -2.0 vdc do. Using an e_c change of 1 volt makes a convenient choice as the divisor for the denominator of the A_V calculation. The change in plate voltage between points "e" and "f" along the AC load line is 64 volts. Dividing this change by the change in e_c gives A_V = 64 / 1 = 64, acceptably close to the specified 60.

So far, we have the self-bias resistor R_2 = 2700 Ω and the plate resistor R_3 = 220 KΩ. To estimate the cathode bypass capacitor, we choose f_{low} = 30 Hz/10 = 3 Hz << 30 Hz, for which the minimum C_1 value is $1 / (2\pi \cdot 3 \cdot 220{,}000)$ = 24 µF.

The voltage drop across is small (2700·0.00052 = 1.4 vdc), so choose an electrolytic capacitor rated at least 10 vdc.

[Parts List Choice: R_2 220 KΩ ½ watt carbon film resistor AES R-A220K]
[Parts List Choice: R_3 2700 KΩ ½ watt carbon film resistor AES R-A2D7K]
[Parts List Choice: C1 47 µF at 16 vdc electrolytic capacitor Jameco 2306279]

Next, choose a value for the C_2, the coupling capacitor. Its minimum value for f_{low} = 30Hz/10 = 3 Hz is $1 / (2\pi \cdot 3 \cdot 690{,}000)$ = 0.008 µF. While a 0.01 µF capacitor would do, using a readily available 0.1 µF here would not be unreasonable. It pushes f_{low} down to 0.2 Hz.

In the old days of paper and tin foil capacitors, large capacitances like 0.1 µF were physically large and, when placed next to the metal chassis, increased the circuit capacitance to ground. This increased capacitance, in turn, affected high-frequency response. Modern capacitors are much smaller, and the detrimental effects are considerably less. For this reason, choosing a coupling capacitor in the 0.1 µF range is reasonable and gives an even better low-frequency response.

As for the 0.1 µF capacitor's working voltage, find the voltage difference between the plate of V_1 (266 − 0.00052·220,000 = 151 vdc) and the control grid of V_2 (0 vdc). It is (151 − 0) = 151 vdc. Choose a capacitor with a rated voltage greater than 151 vdc.

[Parts List Choice: C2 0.1 µF at 630 vdc metal film capacitor AES C-TD1-630]

The last component is the control grid resistor R_1. As the specification calls for an input resistance of 100 KΩ minimum, we choose 1 MΩ.

[Parts List Choice: R_1 1 MΩ ½ watt carbon film resistor AES R-A1M]

The figure below shows the schematic with component values.

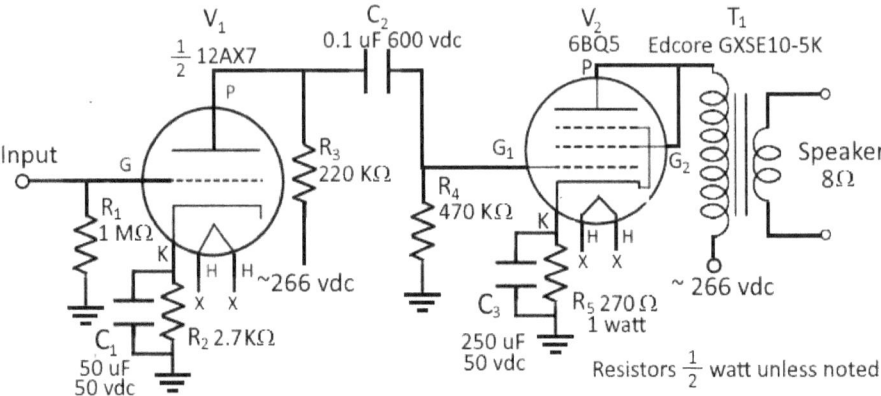

Though the basic SET design is now complete, the 6% distortion question still lingers. In the next chapter, we use breadboard testing to address this and confirm the other operational parameters.

Chapter 10 – SET Breadboard Testing

We have acknowledged all along that the design process is not exact. The theoretical results we derived often relied on assumptions that might or might not be valid under all conditions. Component tolerance is another factor. Most of the components we use have tolerances that range from 5% to 10%. Tube plate characteristics also contribute as they are approximate at best and vary with such factors as the tube's age and manufacturer.

We also have a lingering question related to the level of harmonic distortion our design produces. The design estimate is 6%, not the 4% given in the specification. It could be that the 6BQ5 is a poor choice for our power output tube, and we should look for a higher power tube.

Given all this, committing a design immediately to a printed circuit board is probably not a wise idea. The smart approach is to breadboard the design and test the circuit first. We can check the specifications and tweak component values if necessary.

Breadboarding in the old days often consisted of prototyping a circuit by constructing it on a wooden board intended for cutting bread. Once upon a time, I made one this way with octal sockets and nails driven into the board as tie points. Over the years, I have constructed other breadboards with octal, 7-pin, 8-pin, and 9-pin tube sockets. Rather than wood, I use Plexiglass with 1" by 2" wood strips as feet. I also incorporate a modern plastic breadboard where I do most of my wiring. See the example below.

While not pretty or sophisticated, it works well for audio experimentation. I bring tube pins out to a modern breadboard, where I place the passive components. The banana pins on the left side connect to a variable high-voltage DC power supply that also provides AC heater voltage. A few Fahnestock clips scattered around provide for external connections, such as to an output transformer.

Admittedly, the lengthy wiring and lack of compactness in component placement can affect stray capacitance, which, in turn, can lead to instability and decreased high-frequency response. In most of my experience, instability is rare, and I can accept the effect on high-frequency response, knowing that the final circuit construction decreases stray capacitance.

The figure below shows the test configuration I use.

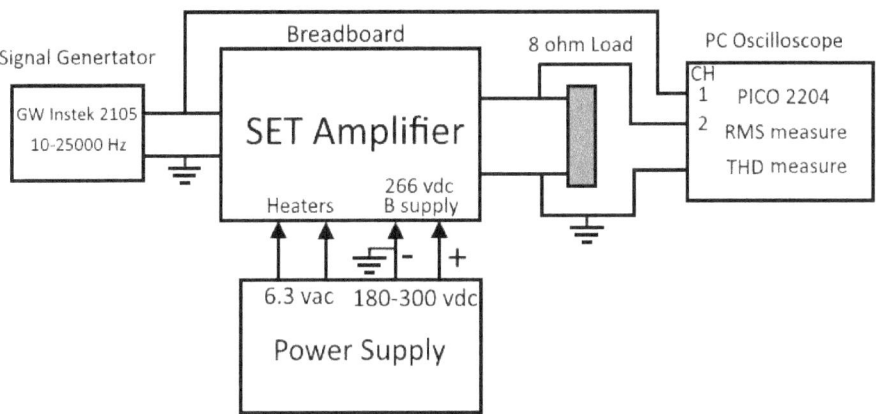

The signal generator is a GW Instek model 2195, which produces an extremely low distortion sine wave over the test range of 10 to 25,000 Hz. The PC oscilloscope is a PICO model 2204 that provides both a visual display of the output and measurements of rms output voltage and total harmonic distortion. The output load is an 8-Ω 5-watt resistor. Lastly, a homebrew power supply provides the 6.3 vac for heaters and variable high voltage DC for B+ supply.

For the test, both the B+ and B++ supplies operated at 266 vdc. The table below gives the test results.

Specification	Design Value	Test Value	Comments
6BQ5 plate voltage	250 vdc	252 vdc	B+ = 266 vdc
6BQ5 plate current	36 ma	31 ma	B+ = 266 vdc
12AX7 plate voltage	148 vdc	150 vdc	B++ = 263 vdc
12AX7 plate current	0.52 ma	0.47 ma	B++ = 263 vdc
12AX7 voltage gain	64	62	1000 Hz & 1 watt
THD at 1000 Hz	6%	3.8%	P_{out} = 1.0 watts
Frequency Response	30-15,000 Hz	38-21,000Hz	-1 db at 1 watt
Input for 1 watt out	0.5 vpp	0.28 vpp	

The values look good, well within the theoretical values. The harmonic distortion number is especially good, below 4% at 1 watt and happily below the 6% predicted value. That is one question that has resolved itself quite satisfactorily!

With confidence in the design, we can proceed to the next chapter with the design of the power supply.

Chapter 11 – Power Supply Design

The purpose of the power supply is to provide AC and DC operating voltage to the voltage amplifier and power output stages of the SET amplifier. In choosing a design, we must consider several factors.

First, we must meet the power voltage and current requirements of the power output and voltage amplifier stages. For our SET design, we need a DC operating voltage (B+) of 266 vdc for both stages. We must allow for the 35-mA current requirement for each power output stage and 1 mA for the voltage amplifier stage. Thus, the minimum DC operating current is 72 mA at 266 vdc.

To accommodate variations in component values, we must allow a 10% overhead in DC operating current. With this in mind, we raise the power supply requirement to $72 + 0.1 \cdot 7.2 \approx 80$ mA at 266 vdc.

Second, power supplies do not produce perfectly steady DC voltage. An alternating voltage component called *ripple* rides on the DC voltage. The ripple for a full-wave power supply is a 120 Hz waveform that can introduce an annoying hum if not addressed.

Because the signal voltage on the plate of the power output stage is large, about 150 to 200 vpp, a power supply ripple of 1-volt peak-to-peak on the B+ supply voltage introduces little hum in the output circuit.

For the voltage amplifier, however, the situation is quite different. The plate signal voltage is less than 20 vpp, so even a 1-volt peak-to-peak ripple is significant and causes an annoying hum. Therefore, the allowable ripple for the voltage amplifier B++ must be much less than that of the power amplifier B+. A reduction factor of 100 or more is usually sufficient.

The two ripple requirements of the B+ and B++ supplies suggest that our power supply design must have two voltage supply points. A higher current supply point for the power output stage with an allowed ripple of, let's say, 1 vpp or so, and a low current supply point for the voltage amplifier stage with a ripple 1/100 as large, say about 0.01 vpp.

Third, we want our power supply to hold a steady voltage as the plate current of the power output stages varies with the audio signal. For our SET design, we must tolerate a current change of up to 100 mA without the output voltage varying significantly. Otherwise, we stand to introduce considerable distortion, especially at low audio frequencies.

Fourth and last, the heater supply must be capable of powering the heaters of the tubes. Generally, we accomplish this with an AC winding on the power transformer. Our SET amplifier voltage requirement is 6.3 vac. Each 6BQ5 requires 0.76 A, and the 12AX7 0.30 A for a total of 2 x 0.76 A + 0.30 A ≈ 1.8 A. Again, allowing for a 10% overhead, the transformer should have a heater winding of 6.3 vac at 1.8 + 0.1 · 1.8 ≈ 2.0 A.

Two power supply types commonly employed in tube-operated equipment are the *capacitor input* and the *inductor/choke input*. Capacitor input designs have greater voltage output for a given input AC voltage. Choke input designs produce lower output voltage but offer better voltage regulation. Given that we are looking for good voltage regulation, the choke input power supply is a better choice for our SET amplifier.

The circuit below shows the choke input design we use in our SET amplifier.

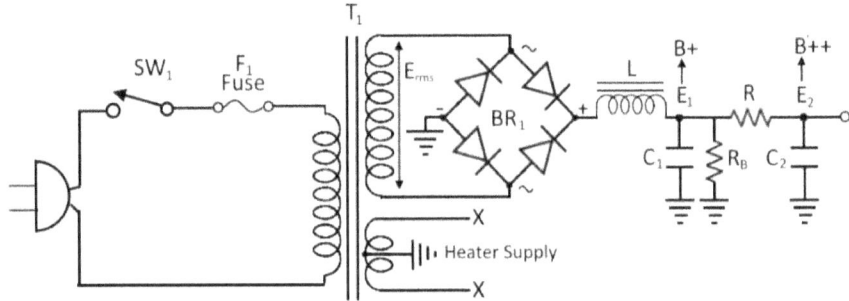

Voltage E_1 with a choke-capacitor filter supplies the power output stage's B+, while E_2 with a resistor-capacitor filter supplies the lower ripple voltage for the voltage amplifier stage's B++.

Let's walk through the design, beginning with the T1 power transformer. T1 transforms the AC line voltage from 120 vac (vac is AC rms volts) into two AC voltages, one a high voltage winding for the tube plate supply and the other a low voltage winding for the tube heater supply.

Commercially available power transformers have a high voltage winding in the 200 to 700 vac range, a low voltage winding of 6.3 vac for the amplifier tubes, and sometimes a 5 vac winding for a rectifier tube. Current handling capacities vary widely.

In addition to providing the AC voltages, an important function of the power transformer is that it isolates the amplifier circuitry from the AC power line. Since there is no direct connection between the power line and the secondary of the transformer, we can safely connect the power supply ground (the DC negative) to the metal chassis. In addition, grounding the chassis helps reduce AC-induced hum.

An important feature of the power transformer is that it transforms these AC voltages with little energy loss. A convenient way of stating this is that the power input to the primary winding nearly equals the power delivered by the secondary windings. The difference is a small energy loss as heat. We use this fact presently in carrying out our design.

The transformer T1 primary connects to the power line via a suitably sized fuse F1 and an SPST switch SW1. The rule of thumb is to select a fuse rated at least two times the normal operational current in the transformer primary. At turn-on, a surge of current much larger than the normal operating current can occur. It lasts only a fraction of a second, but that is enough time to blow the fuse specified this way. To avoid this possibility, we use a Slow-blowing fuse specially designed to handle surges yet blows if excessive current persists for more than a brief time.

To obtain the primary current, we can use the transformer's power-in equal power-out feature mentioned earlier. To calculate the secondary power delivered, multiply each winding voltage by its current rating, then add the results. Call this P_S, which is the total secondary power delivered in watts. If the power input to the primary is P_P at 120 vac and assuming $P_p = P_s$, the fuse current would be twice $P_P / 120$.

One caution is that the fuse current should never be more than the worst-case power transformer current calculated by dividing the AC line voltage by the DC resistance of the transformer primary. While this is not usually a problem, we should check it. However, to check it, we must wait until we select a transformer and know its primary winding resistance.

For our SET amplifier, the heater current at 6.3 vac is 1.8 A (2 x 0.76 A + 0.30 A). The heater power is then 1.8 A x 6.3 vac = 11.5 watts. The plate supply power is 0.072 A x 266 vdc = 19.1 watts. Therefore, P_s is 11.5 watts + 19.1 watts = 30.6 watts. The primary current is 30.6 watts / 120 vac = 0.26 A. A 0.5 A Slow-Blow fuse is a desirable choice for F_1.

[Parts List Choice: F1 – 1/2 A 250 volt Slow-Blow Fuse AES F-Z3AG-S0D5]

Choosing the on-off switch SW_1 is not difficult. We must be sure it has the voltage and current rating for our application. A voltage rating of 120 vac and a current rating of 1 A or greater does nicely.

[Parts List Choice: SW1 – 3 A 125 SPST ON-OFF Switch AES 002574]

We can now select a transformer. The main considerations here are the voltage and current of the high voltage winding and the current of the 6.3 vac heater winding. Previously, we estimated the current requirement as ~2 A.

To specify the high voltage winding, we start with the specified DC operating voltage and current requirement. For our SET amplifier, the target E_1 is 266 vdc, and the current is 72 ma.

For a choke input filter, voltage E_1 is

$$E_1 = \frac{2\sqrt{2} \cdot E_{rms}}{\pi} - I_L \cdot R_{CH}$$

E_{rms} is the secondary voltage of the transformer. Variable R_{CH} is the DC resistance of the choke, and I_L is the choke current. The formula neglects the forward resistance of the bridge diodes as it is minor compared to the R_{CH}.

Solving for E_{rms}, we get

$$E_{rms} = \frac{\pi}{2\sqrt{2}} (E_1 + I_L \cdot R_{CH}) = 1.11 \cdot (E_1 + I_L \cdot R_{CH})$$

Since we haven't chosen a choke, we must assume a value for R_{CH}. A check of several chokes suggests that a DC resistance of 125 Ω would be about right. If, later, the value for the choke we choose is significantly different, we can always recalculate. E_1 is 266 vdc, and I_L is 72 ma, the DC operating current.

Substituting in the formula above, we calculate E_{rms} to be 304 vac. Applying the 10% overhead to the 72 mA current requirement, we need to look for a power transformer with a high voltage winding of ~305 vac at ~80 mA and a low voltage winding of 6.3 vac at least ~2 A.

Fortunately for us, the Hammond 268BX power transformer closely fits the bill with 300 vac at 86 mA and 6.3 vac at 2 A.

[Parts List Choice: T1 – 300 vac at 86 mA and 6.3 vac at 2 A AES P-T269AX]

As a further note, let's consider what we would do if the exact high-voltage winding were not available. Choosing a lower voltage secondary would reduce the SET's power output and might be an acceptable solution. We could switch to a capacitor input power supply that would raise the output voltage at the expense of voltage regulation. The problem here is whether we are willing to accept increased distortion at low audio frequencies.

Choosing a higher voltage secondary is also a possibility. In this case, we could add a resistor between the bridge and inductor to introduce a voltage back to 300 vdc. The added resistance would worsen voltage regulation. So long as the inserted value is small, less than 200 Ω, we might tolerate the effect on voltage regulation. To add this resistor, we must size it for the correct power handling capacity. Using the $I^2 \cdot R$ formula, a 200-Ω resistor would need to be at least $0.072^2 \cdot 200 \approx 1.0$ watts. A 2-watt resistor would do.

Now that we have selected a power transformer, we can check for worst-case current. Recall that this is line voltage / primary DC resistance. For the selected transformer, the primary DC resistance is 13 Ω. The worst-case current is then 120 / 13 = 9 A. The chosen fuse size of 1/2 A is well below this.

Next is the bridge rectifier. The bridge rectifier must have a current handling capacity of at least twice the DC operating current and a peak reverse voltage greater than the DC operating voltage. Given our 266 vdc operating voltage and 75 mA current, we chose a 2 A 400 peak-reverse-voltage bridge rectifier.

[Parts List Choice: BR1 – 3 A 400 PIV Bridge Rectifier AES P-QBR-34]

The voltage appearing at the bridge output is a full-wave sinusoid with a peak value equal to $\sqrt{2} \cdot E_{rms}$. See the figure below.

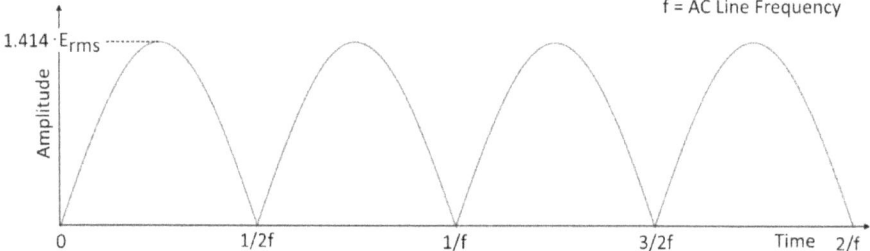

We need a smoothing filter to fill in the zero-going dips between peaks. As noted above, the choke input filter not only smooths the peaks but also offers highly desirable voltage regulation.

As shown in the power supply diagram above, the choke input filter consists of choke L and electrolytic capacitor C_1. E_1 is the voltage source for the power output stage B+.

We first calculate the choke filter components, L and C_1. The size of these components depends on how much ripple voltage (call it ER1) is acceptable in E_1. We can estimate the product of L and C_1 with this formula.

$$L \cdot C_1 \cong \frac{4}{3\pi} \cdot \frac{\sqrt{2} \cdot E_{rms}}{4 \cdot (2\pi f)^2 \cdot E_{R1}} = 1.055 \cdot 10^{-6} \cdot \frac{E_{RMS}}{E_{R1}} \; (for \; f = 60 \; Hz)$$

E_{R1} is the allowed ripple voltage, which we limit to 1 vpp. For our SET amplifier, E_{R1} = 1 vpp and E_{rms} = 300 vac. Then, $L \cdot C_1$ = $3.165 \cdot 10^{-4}$. (Note that "vac" is the same as volts "rms.")

As a practical matter, it is best to choose from readily available chokes first and then calculate the value of C_1. If C_1 is too large, then choose a choke with a larger inductance. We should choose one with a current rating at least 10% above the DC operating current. In our case, 110% of 72 mA is 80 mA minimum.

Chokes typically come in multiples of 5 H, so we start with a 5 H choke. We find that C_1 is $3.165 \cdot 10^{-4}$ / 5 or 63 µF minimum, which is quite reasonable. So, for our SET amplifier, we choose a readily available choke with L = 5H at 120 mA and 100 uF electrolytic capacitor.

The capacitor's operating voltage should be at least 10% greater than the transformer's peak high voltage E_p given below.

$$E_p = \sqrt{2} \cdot E_{rms} = 1.414 \cdot 300 = 424 \; vdc$$

A 450 vdc electrolytic capacitor does.

[Parts List Choice: L – 5 H at 120 mA AES P-C194F]
[Parts List Choice: C1 –100 uF at 450 vdc AES C-ET100-450]

The second stage filter to supply the lower ripple B++ uses a series resistor R and a second electrolytic capacitor C_2 for additional smoothing. We want a reduction of the ripple at E_1 by a factor K_{RC} = 100. Therefore, we use the formula below to determine the value of $R \cdot C_2$.

$$R \cdot C_2 = \frac{K_{RC}}{4\pi \cdot f} = \frac{100}{4\pi \cdot f} = \frac{7.96}{f} = 0.133 \; (for \; f = 60 \; Hz)$$

Because E_2 supplies only the voltage amplifier's small current (a few milliamps at most), a 47 uF capacitor is large enough for our SET amplifier. Solving for R in the formula above, we get R = 0.133 / 47×10^{-6} = 2800 Ω. The nearest 5% standard value, 2700 Ω, does fine.

[Parts List Choice: C1 – 47 uF at 450 vdc AES C-ET47-450]
[Parts List Choice: R – 2700 Ω ½ watt AES R-A2D7K]

Finally, we add *bleeder resistor* R_B. It serves primarily as a safety measure to discharge the electrolytic capacitors when no load is present. Without the bleeder resistor, the capacitors could carry a high voltage charge for several minutes, presenting a danger to a person who is contacting the circuit.

The choice of bleeder resistor size is a tradeoff between discharge time and wasted power. A smaller resistor discharges the capacitors more quickly but wastes power. A larger resistor wastes less power but leaves the capacitor voltage at dangerous levels longer. A discharge time of 15 seconds is a reasonable compromise.

Discharge time t_d is the product of the bleeder resistance and total capacitance. So, $R_B = t_d /(C_1 + C_2)$. If we use microfarads for capacitance, then R_B is in megaΩ.

For our SET amplifier, $C_1 + C_2 = 147$ uF. Assuming a discharge time of 15 seconds, we calculate $R_B = 15 / 147 \cdot 10^{-6} = 102{,}000$ or, choosing a standard value of 100 KΩ. To obtain the resistor's power rating, we compute $P = E_1^2 / R_B = 0.707$ watts, so we need a 1-watt resistor.

[Parts List Choice: R_B –100 KΩ 1 watt AES R-B100K]

Warning! When examining a circuit containing high voltages, sound practice is to intentionally discharge capacitors manually whether a bleeder resistor is in place or not!

The last step is to check that the choke DC resistance is close to the 125 Ω we used in the transformer calculation. Checking the P-C194F specification sheet, we find its DC resistance is 115 Ω, close enough not to require recalculation.

Here is the complete power supply.

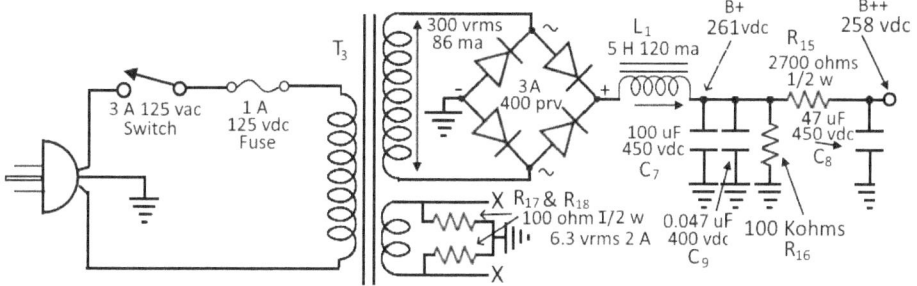

The lower B+ and B++ consider the lower transformer high voltage (300 vac, not 304 vac). Note also that the heater winding of the transformer is not center-tapped. It was then necessary to place the series 100-Ω resistors across the heater winding and ground the common point as shown.

[Parts List Choice: 2 x 100 Ω ½ watt AES R-A100]

With the power supply completed, we have only to address a few minor modifications to the circuit before committing to a build. In the next chapter, we make a few refinements to our design.

Chapter 12 – Circuit Refinements

Now that we have completed the basic design, we should consider a few refinements before committing to a printed circuit board. Here is the basic SET design.

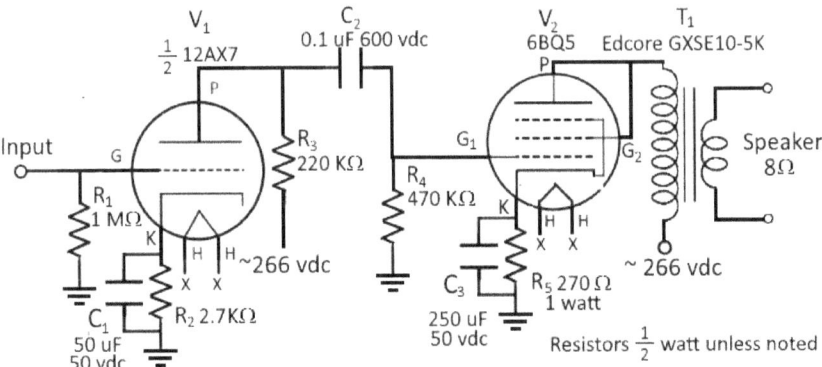

Add Volume Control

Because we do not want to always operate at maximum gain, we add a potentiometer in place of resistor R_1 to act as a volume control. See the modified schematic below.

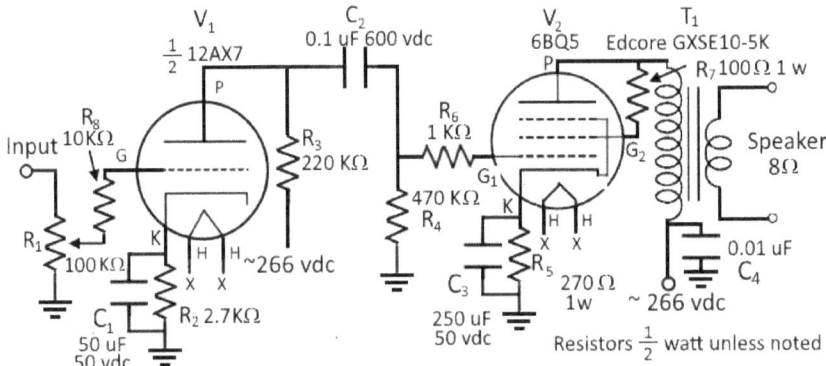

The volume control wiper connects to the control grid of the voltage amplifier so that the input always presents a constant load to the input source. Otherwise, the varying load as we adjust the control could change the operating characteristics of the input source.

To exceed the 100 KΩ input specification, we first selected a 1 MΩ potentiometer, thinking the higher, the better. Doing so turned out not to be a good idea.

One of the weaknesses of triode amplifiers is the multiplying effect of interelectrode capacitance called the *Miller Effect*. True, the interelectrode capacitances are quite small, a few picofarads. However, when we include the Miller Effect, the stage gain (60 in our design) multiplies this small capacitance. See Appendix G for more on the Miller Effect.

The capacitive load at the control grid of the 12AX7 is approximately 60 times 3 pF = 180 pf. Add the capacitance of the tube socket and wiring, and we have up to 200 pf or more. When we introduce a volume control, we are, in effect, adding a low pass filter made up of the volume control resistance and Miller Effect capacitance. See the figure below.

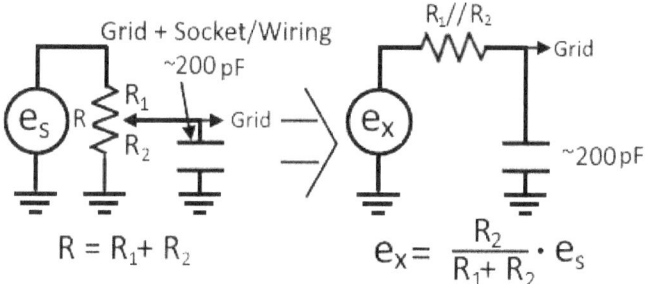

On the left is the actual circuit. On the right is the Thevenin equivalent, showing the variable input voltage (the volume control), the series Thevenin resistance, and the ~200 pF grid capacitance. It is easy to see that the capacitance bypasses higher frequencies to the DC ground, possibly affecting the high-frequency response of the amplifier.

The low frequency breakpoint f_{low} = 1 / ($2\pi(R_1 \cdot R_2/(R_1+R_2)) \cdot 100 \cdot 10^{-12}$). For the potentiometer, the maximum resistance occurs when $R_1 = R_2$. For a 1 MΩ potentiometer, we have 500 KΩ // 500 KΩ and f_{low} = 3183 Hz, which would seriously affect the amplifier's high-frequency response.

While the 1 MΩ volume control sounded like a promising idea, we would be introducing a serious frequency response issue. Therefore, we must reduce the resistance, but going too far would lower the input resistance and load down input devices connected to the amplifier. Selecting a 100 KΩ potentiometer is a good compromise, raising f_{low} to 32 KHz, taking it out of the 20 kHz response limit of the amplifier while still meeting the minimum input load specification.

[Parts List Choice: R1 – Dual 100 KΩ ½ watt pot-audio taper AES R-V38-2X100KA]

Add Stopper Resistors

Occasionally, parasitic oscillations occur in SET amplifiers, so taking precautions to eliminate the possibility is a good idea.

The power output tube is a likely place in our design for instability to occur. Specifically, the control and screen grids are the places where unwanted feedback can produce intermittent or continuous high-frequency oscillation. So, it is where we apply our cautionary remedies.

Since the cause is feedback via the control and screen grids, what we need is a low pass filter that permits audio frequencies to pass unhindered and traps out high frequencies that contribute to parasitic oscillation.

In circuit design, a simple low-pass filter consists of a connected resistor and capacitor, as shown below.

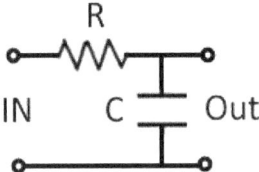

The idea is that the capacitor is an open circuit for audio frequencies and a short circuit at the higher frequencies, causing parasitic oscillation. For this reason, we refer to R as a *stopper resistor*. The breakpoint frequency at which the capacitor's impedance equals R is $f_{low} = 1 / 2\pi RC$.

To be effective, we must place resistor R as close as possible to the grid, which needs low pass filtering. In the schematic diagram above, R_6, in conjunction with the power output tube's interelectrode capacitance, performs the low pass function. We call R_6 the grid stopper resistor. For power output triodes, a typical value for the grid stopper resister is 1 to 10 KΩ. In this size range, no appreciable attenuation occurs at audio frequencies. Since no appreciable current flows in the grid stopper resistor, a ½ watt resistor does fine.

[Parts List Choice: R6 – 1000 Ω ½ watt AES R-A2D1K]

R_7 is the stopper resistor for the screen grid. Values here are much smaller; 100 Ω is a common value. To accommodate the screen current in R_7, a 1-watt resistor should be used. Rather than increasing resistance if oscillation persists, try substituting a small inductor of a few millihenries.

[Parts List Choice: R7 –100 Ω 1 watt AES R-B100]

Because the volume control wiring to the voltage amplifier control grid is necessarily lengthy, it is a good idea to use a 10 KΩ stopper resister there for go measure.

[Parts List Choice: R8 – 10000 Ω ½ watt AES R-A2D10K]

As we have noted, for the stopper resister to be effective, it must be adjacent to the grid it is protecting. Taking this into account while laying out the wiring is important.

Add Audio Bypass Capacitor to Power Supply

The last refinement is a 0.047 µF capacitor connected from the B+ supply to the DC ground. See below.

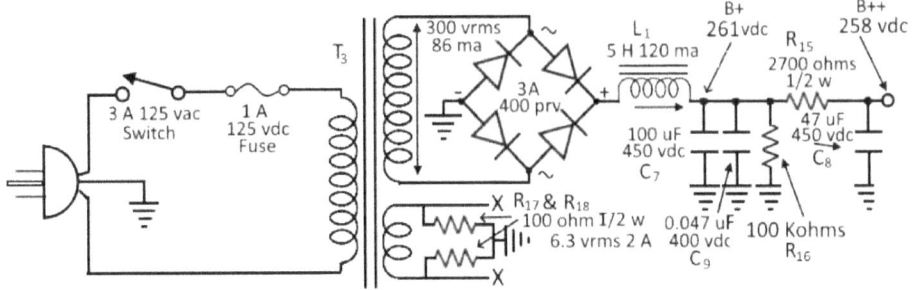

Recall that we assumed that the B supply was a short circuit at audio frequencies. We could assume that the large electrolytic capacitors in the power supply perform this task, and they do so for low audio frequencies. However, as an added measure to cover higher audio frequencies, we include a modest-sized conventional capacitor between the B+ and ground.

[Parts List Choice: C4 – 0.047 µF at 400 vdc AES C-PD047-400]

With these refinements completed, we can move on to the design of a printed circuit board in the next chapter.

Chapter 13 – Printed Circuit Board Design

While hand-wiring our SET amplifier is certainly an option, designing and using a printed circuit board makes construction much easier. In this chapter, we go through some basic steps in PCB design and finish with the information needed to produce a printed circuit version of our SET amplifier.

Electrical Safety

At the outset, I must stress the inherent dangers of the high voltages in tube amplifiers. **These voltages are potentially fatal! You must always exercise care to avoid contact with circuits alive with these voltages.** Here are some safety rules to follow:

* Never work on an electronic circuit with power applied.
* Treat all electronic circuits as energized, whether they are or are not.
* After removing power, safely discharge all electrolytic capacitors.
* Keep yourself and the work area dry.

* Avoid contact with any earth-grounded object.
* Never use metallic tools or items to probe an electrical circuit.

Finally, always follow the *One Hand Rule* that states, "If it is safe to do so, work with only one hand, keeping the other hand at your side or in your pocket, away from all conductive material." The greatest danger is an electrical current passing through the chest and upsetting the heart's rhythmic beat.

If you are unable or unwilling to follow these rules, then do not attempt to construct this SET amplifier.

Printed Circuit Board Sources

Several internet companies provide PCB services, from schematic capture (creating a schematic) to producing a finished printed circuit board. I use ExpressPCB (https://www.expresspcb.com/). They provide free software to create a schematic, lay out the PCB design, and order printed circuit boards. The discussion that follows utilizes ExpressPCB software.

PCB Design Steps

The first step is to create a complete schematic of the SET amplifier. We have already developed the basic design, including the power supply. The only addition is making allowance for two amplifier stages.

Several videos are available on the Internet to introduce how to use ExpressPCB schematic capture software. I like this one:
https://www.youtube.com/watch?v=0SQnGvI2ang.

Appendix K shows the complete schematic as entered with ExpressPCB software. Once we create the schematic, the next step is to design the printed circuit board. Below is the SET printed circuit board, followed by a discussion of various aspects of the PCB design process.

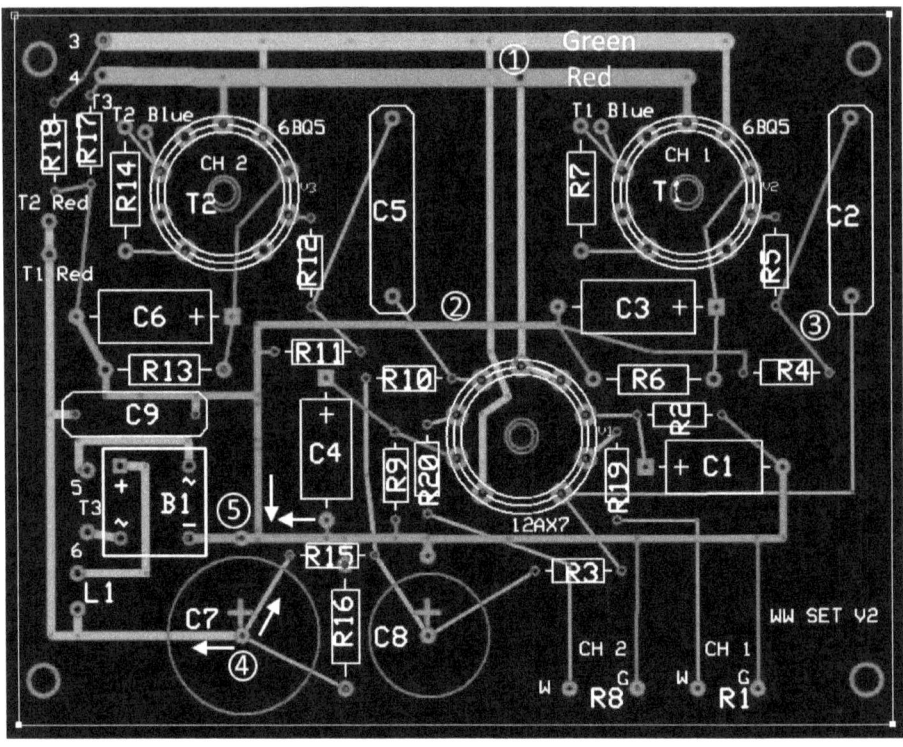

Routing Tips

Instead of wires the PCB uses thin copper lands on an insulating sheet. A *two-layer board* has *lands* on the top (labeled Red in the figure) and bottom (labeled Green in the figure). Round pads pierce the board and connect lands on one side of the board to those on the other. For more complex designs, multilayer boards are available with lands sandwiched between insulating sheets. Up to four layers are available at ExpressPCB for more complex designs. For our SET design, we need only two layers.

We deal with varying current carrying capacities by selecting the width of the land. Here are the ExpressPCB recommendations.

0.010" 0.3 Amps
0.015" 0.4 Amps
0.020" 0.7 Amps
0.025" 1.0 Amps

0.050" 2.0 Amps
0.100" 4.0 Amps
0.150" 6.0 Amps

In general, 0.15" to 0.025" are good sizes for signal and low current circuits, while 0.050" to 0.100" lands are more suitable for DC supply and high current AC heater circuits.

Note that the bridge rectifier's AC inputs with squiggly lines ~ connect to T1's high-voltage winding. Plus, "+" and minus "-" signs mark the DC outputs. These marks match similar marks on the bridge rectifier to assist in the correct orientation when mounting the device.

For the PCB design above, B+ and ground lands are 0.050" (point 1 in the figure above), 6.3 vac heater lands are 0.100" (point 2), and signal lands are 0.015 or 0.020 (point 3). Note that we routed high-voltage AC and DC lands on the underside of the board for safety reasons. When exposed during testing, the board's dangerous voltages are out of reach on the bottom of the board.

In general, try to locate each tube and its associated components close together and keep interconnecting lands as short as possible. In the case of a stopper resistor, keep one end immediately next to the tube's pin. See R19 and R20 in the PCB image above.

When laying out B+ supply and ground lands, always route them from each stage to single B+ and ground points (points 4 and 5 above, respectively). Doing so prevents currents from intermingling from various stages, thus creating unwanted feedback paths and sources of AC hum. Designers call this a "Star" wiring strategy.

A related strategy specifies that the power supply ground should connect to the metal chassis at a single point. For this design, make the connection at the RCA jack ground terminals. This arrangement ensures that no currents and voltages flow in the chassis that could inject feedback or AC hum into sensitive audio inputs.

As noted earlier, grid stopper resistors must be located immediately adjacent to the grid pin. For example, note the position of the stopper resistor R20 at the control grid of the 12AX7 voltage amplifier.

External Component Connections

External components like the transformers and chokes connect to the board via pads labeled to reflect the component's pin numbers. For more convenient servicing, we could use plugs and sockets, but bringing all the connections to a single area of the PCB would significantly complicate routing. I opted for simplicity!

Schematic and PCB Files

You can find the schematic and PCB files needed to order a printed circuit board from ExpressPCB at www.github.com/rbwhipple/SET_Book. Ordering a single PCB is expensive, but a PCB saves time, makes for a neater appearance, and generally leads to a more pleasant DIY experience.

In the next chapter, we describe the layout of parts on a metal chassis and provide some construction hints.

Chapter 14 – Chassis Layout, Construction and Testing

Using a printed circuit board eliminates much of the wiring, but it doesn't shortcut the more difficult part of the construction, the metal working associated with the chassis. Having the right tools makes it easier. Learning to take your time and not get in a hurry helps, too.

Nevertheless, to build a SET amplifier, a nice metal chassis is almost a necessity. Shielding the dangerous, high voltages from prying hands is reason enough. The metal chassis also reduces noise, particularly power line hum. In this chapter, we cover the basics of chassis construction, but like other skill-based tasks, experience is the best teacher. So, let's begin.

Two types of metal chassis are readily available: aluminum and steel. Aluminum is easier to work with and probably should be the first choice for a newcomer. With experience, steel makes a better-looking and studier chassis.

To determine the chassis' overall size, place the major components in an uncrowded configuration. From this, we can get an idea of how much top surface space we need. The figure below shows a typical arrangement for a SET amplifier.

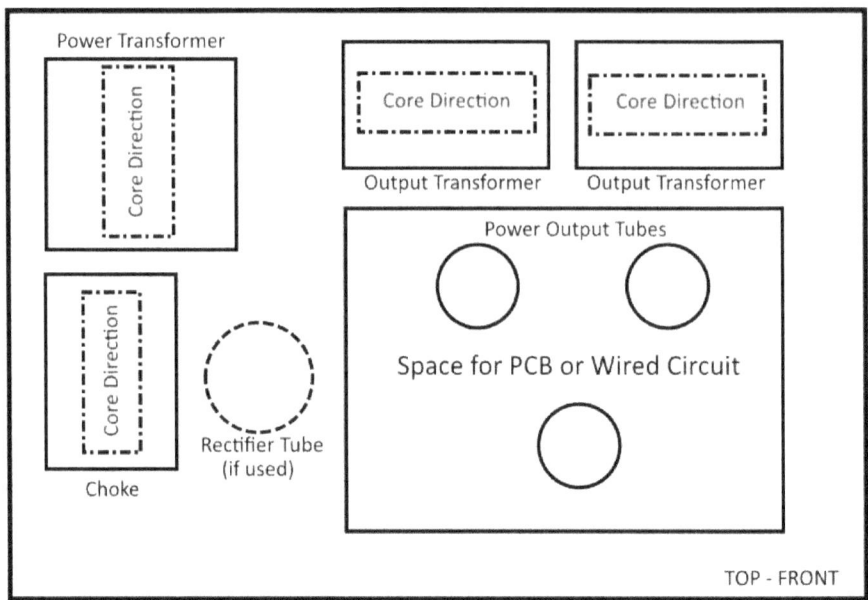

Here are a few things to note.

1. Locate the power supply and amplifier in separate sections as shown. Doing so keeps the AC line voltages away from sensitive audio circuits. Reverse the positions of the choke and power transformer if desired.
2. If using a rectifier tube, the space next to the choke is generally available.
3. Orient the power transformer and choke cores perpendicular to the audio output transformer cores. Doing so reduces the possibility of AC line hum induced by the former into the latter.
4. Locate the audio output transformers toward the rear of the chassis near the speaker terminals on the back of the chassis.
5. Position the power output tubes near the audio output transformers and the voltage amplifier tube close to the volume control at the front.

For the typical SET amplifier, a chassis size of 12" by 8" by 2" works out to be about right. The black steel chassis below is a desirable choice.

[Parts List Choice: Hammond Black Steel Chassis 8"x12"x2" AES P-H1441-22BK3]

To avoid scratching the chassis, cover the top, front, and back with 4 or 8-square-to-the-inch graph paper. Align the graph paper lines with the edge of the chassis. Use rubber cement to affix the paper to the chassis. Since the chassis length is longer than the graph paper, we need to add extension strips of graph paper.

Mark the location and size of holes on the graph paper, which permits drilling and punching holes without marring the chassis. When finished, remove the remaining graph paper and residual rubber cement.

The diagrams below show the hole pattern for the top, front, and back of our SET amplifier.

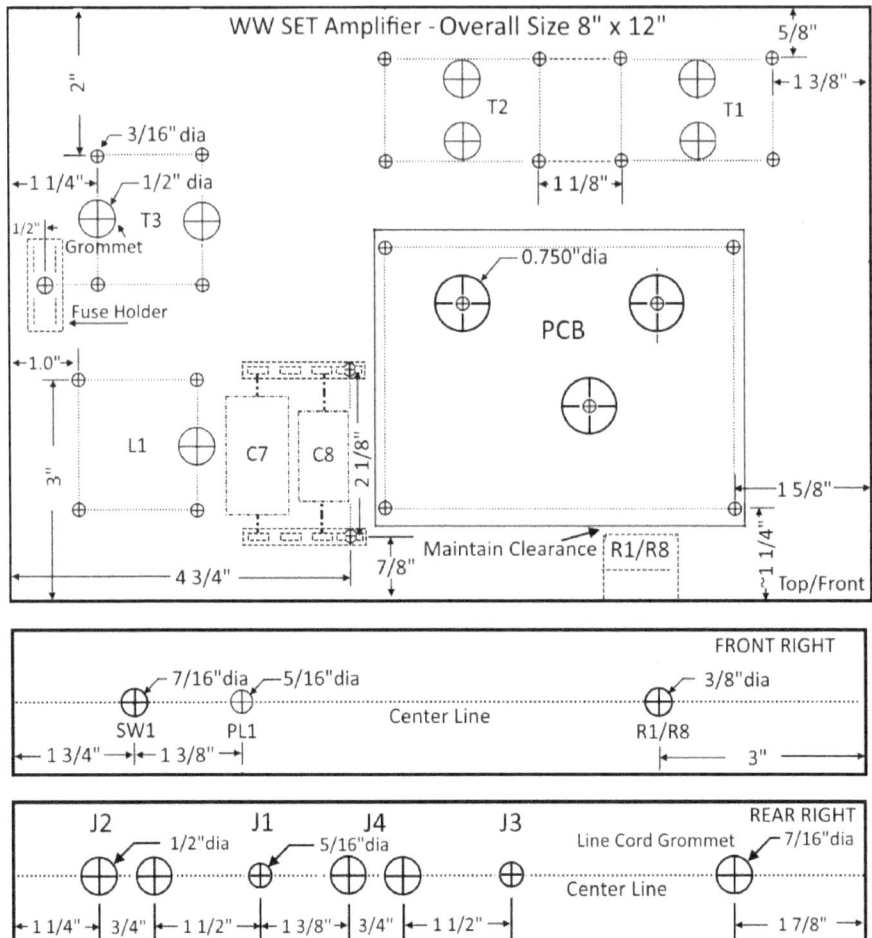

With multiple hole components, it is best to locate and mark one hole and then use the actual component to mark the remaining hole(s). The diagram gives the location of one hole of each multiple-hole component. Place the component aligning this hole, then mark the remaining holes.

For the transformers and choke, position them on the marked hole. Make sure it is parallel to the edge of the chassis and oriented with the wires above the indicated holes. Mark the remaining holes. For each, mark the 1/2" grommet holes (for the wires) in the relative positions shown.

I originally designed the PCB to mount the electrolytic capacitors on the board. Later, to avoid safety issues, I decided to mount the PCB below the chassis with holes in the chassis so the tubes could protrude through the chassis. I moved the electrolytic capacitors below the chassis, as shown in the diagram.

Mark the mounting holes for the two 5-lug terminal strips.

To mark the mounting holes for the PCB, follow these steps using an unpopulated PCB. Position the PCB component side up on top of the chassis, aligning the front and right edges as shown. **The front edge of the PCB must be 1" from the front edge of the chassis to accommodate the dual potentiometer R1/R8.** With the PCB edge parallel to the chassis, mark the four mounting holes and the holes in the center of each tube socket.

Next, mark the holes on the front and rear of the chassis using the dimensions shown.

To drill clean holes, we need the right tools. Because the chassis is steel, quality cobalt drill bits are a worthwhile investment. While a hand drill suffices, the drill press works best. For each hole, use a hammer and center punch to make a small dimple to give the bit a place to start.

Always use safety glasses, and don't get in a hurry. Start by first drilling a hole that is a bit one-half the size of the final hole. Keep a small puddle of 3-in-1 oil around the bit. Use a slow drill speed and only moderate pressure. Be especially careful to take it easy as the drill begins to break through. Use light and pulsing pressure during this final stage. When finished with the smaller hole, drill the specified size hole.

Finally, use a deburring tool to clean up the edges of the hole. The one pictured below by AFA Tooling is an excellent choice and is available from Amazon (search "AFA Tooling Deburring Tool").

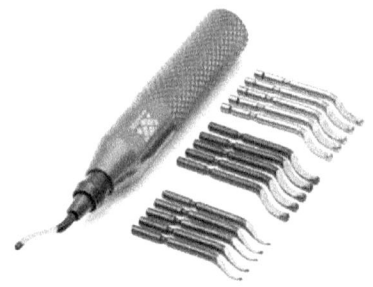

Check out the videos on YouTube to learn how to use the deburring tool.

For example, for the 3/16" holes, use a 3/32" bit first. For large holes, like the 1/2" grommet holes, use two or more smaller bits starting with 1/4". Time spent stage-drilling is time-saved deburring!

For the tube holes, a 3/4" chassis punch is a necessity. Note that 3/4" is the final hole size, not the conduit size, sometimes specified. When purchasing a chassis punch, be sure that it is suitable for steel 0.036" thick.

To use the chassis punch, first drill a 3/8" hole. Accuracy here is important because a small error might create problems later when inserting the tube through the chassis opening. Place the die (the flat-surfaced piece) on top of the chassis and the punch (the sharp, cutting piece) on the bottom. Pass the bolt through the die, then the hole, and lastly, thread it into the punch. As the bolt tightens, the punch cuts its way cleanly through the chassis, leaving the 3/4" hole.

After drilling and cleaning up the top chassis holes, follow the same procedure for the front and rear chassis holes. The dual banana binding posts have 1/2" insets that require holes slightly larger than 1/2". Rather than using a larger bit, use the deburring tool to enlarge the hole. After deburring, continue to ream out the hole until the insets meet and the binding post is securely in place.

Using the deburring tool to increase hole size could apply elsewhere if a drilled hole is slightly small. Remember that a hole that is too small is preferable to one that is too large. We can always enlarge the hole if necessary!

When you have drilled and punched all holes, the next step is to mount the components. Mounting fasteners are 6-32 by 3/8" machine screws. Use washers on top of the chassis and lock washers below. **You need additional washers when mounting the fuse holder to allow the fuse to fit correctly.**

When installing the RCA input jacks J1 and J3, scrape the inside of the rear chassis so that the ground lugs make a good connection to the chassis ground. For our grounding system, these are the only points where the signal/DC ground connects to the metal chassis.

Install the 3/8" grommets in the seven 1/2" holes on the chassis top. Install the 1/4" grommet in the 7/16" hole on the chassis rear. This smaller grommet helps provide strain relief for the line cord.

When populated and before mounting the PCB, use a "White Sharpie Extra Fine Point Oil Based Paint Marker" to identify the external wiring points on the bottom of the PCB. We cannot label the pads once we mount the PCB below the chassis. Figure 8 shows example markings.

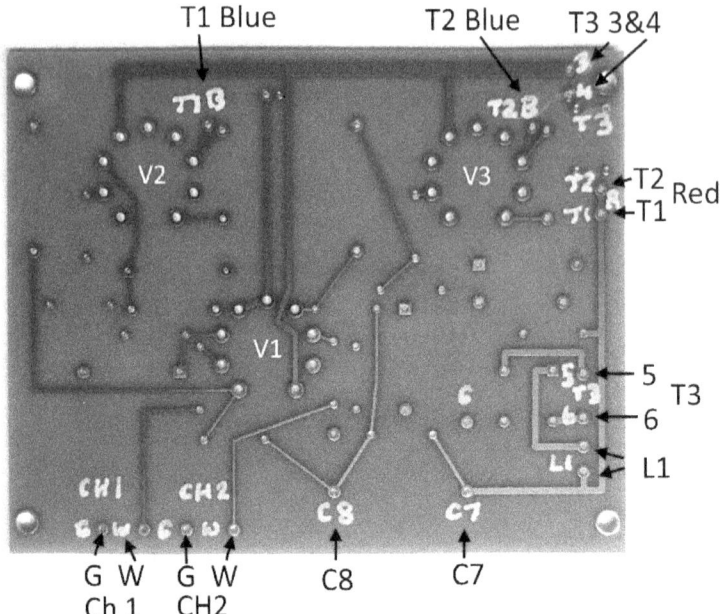

Next, mount the PCB below the chassis using four 6-32 x ½" male standoffs, as shown below.

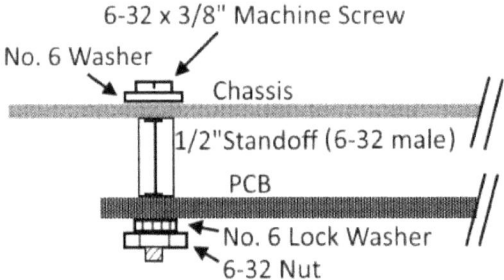

Orient the board so that tubes V2 and V3 are nearest the rear of the chassis, and the top of the PCB faces the chassis from below.

Wire external components to the PCB in stages. In stage 1, wire the electrolytic capacitors. First, install the two 5-lug terminal strips oriented as shown below.

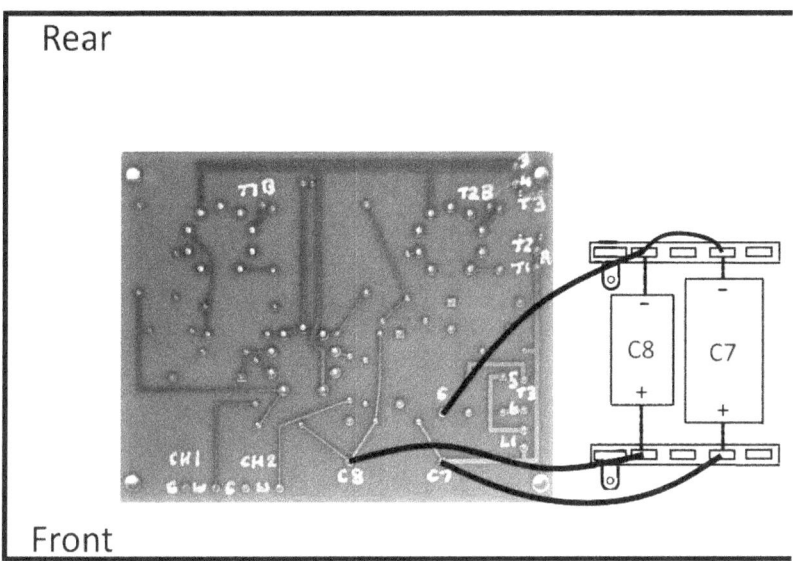

Take care to orient the capacitors as shown with the positive terminals nearest the front of the chassis. Wire the capacitors to the PCB as shown. At the connection to the PCB, use only enough bare wire to enter the hole and then solder the wire. See below.

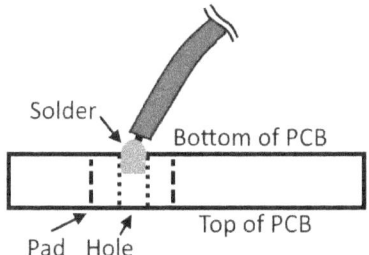

Ensure that no wire protrudes to the top of the PCB, as this could result in a short circuit!

Note that the capacitor negative terminals connect to the DC negative via the hole marked "G" in the figure. **Make no connection to the ground terminals of the terminal strip.**

Stage 2 is the wiring of the transformers and choke. See the figure below.

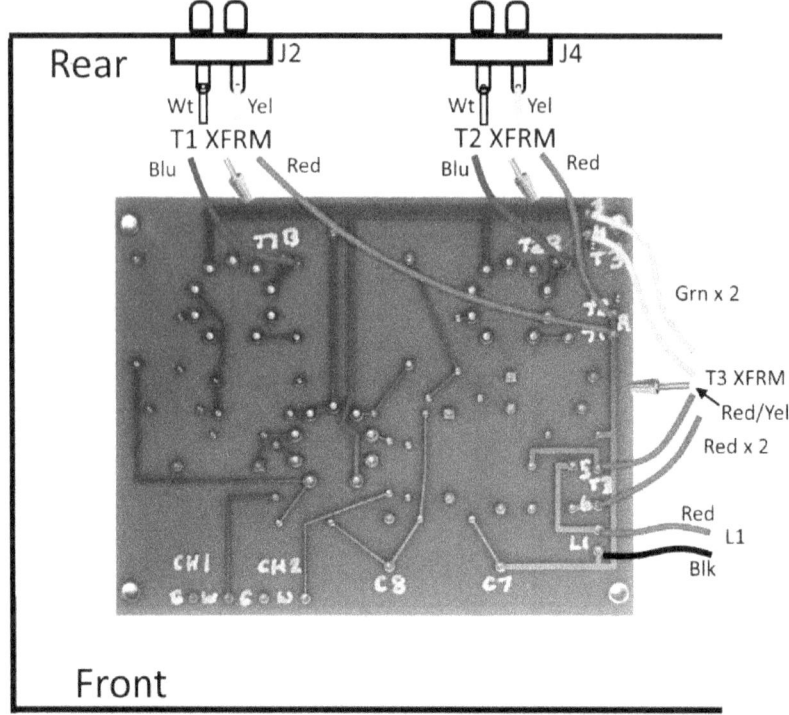

Keep the leads as short as possible while routing them as shown. The blue/white wires from output transformers T1 and T2 are screen taps in ultra-linear designs and are unused. Roll the wire up neatly and cap the end with a wire nut as shown.

Do the same for the red/yellow wire from power transformer T3. This wire is the center tap of the 300-vac secondary and is unused in our design.

Stage 3 is wiring the volume control to the PCB. See the figure below.

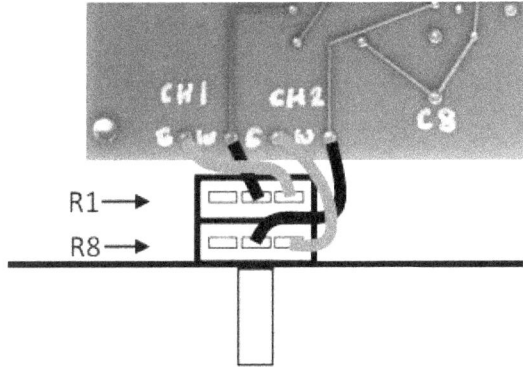

The wiper is the center terminal of the volume control and connects to the "W" point on the PCB. The right-most terminal (as seen from above) connects to the "G" ground point on the PCB. Note that when we turn the volume control fully counterclockwise, the wiper contacts the DC ground, establishing the minimum volume setting.

Stage 4 connects input J1 to the volume control with a length of shielded audio cable, as shown below.

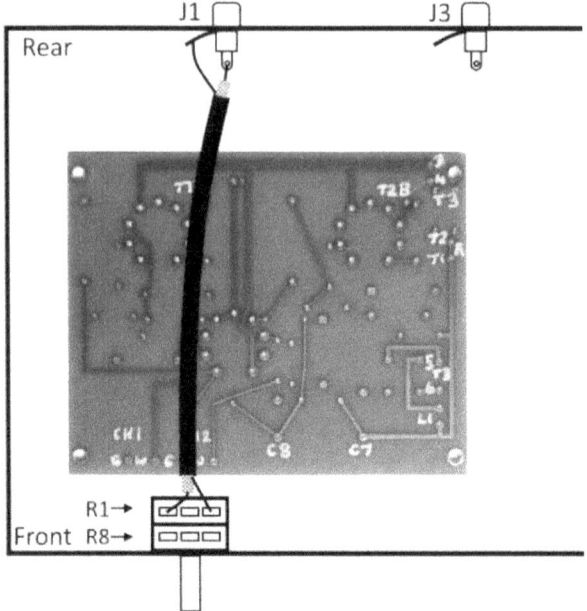

Rather than purchase a shielded cable, it is simpler to sacrifice an inexpensive audio cable by cutting the length needed from between the end plugs/jacks.

Stage 5 is like Stage 4, except the connection is to audio input J3. See below.

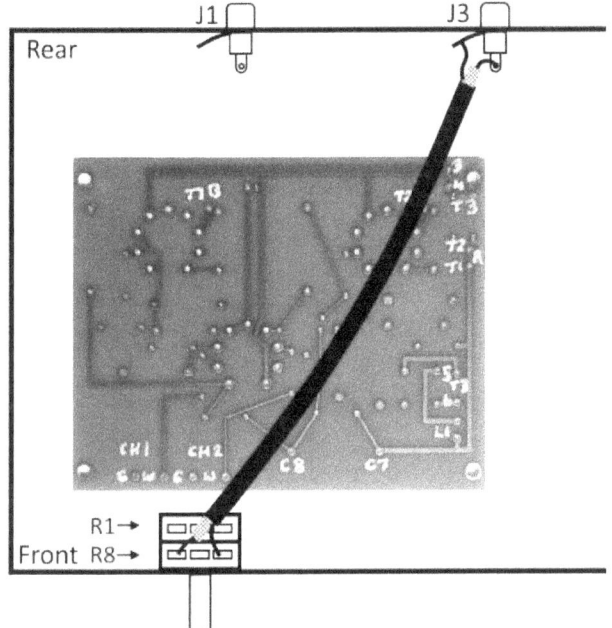

The last wiring is for AC power. See the figure below.

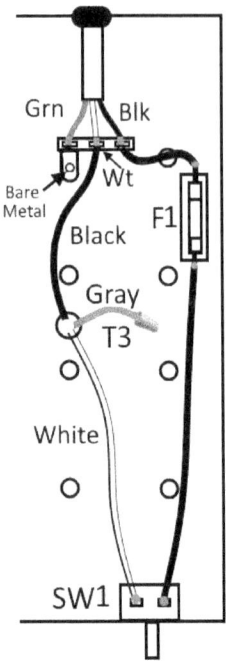

Scrape the paint away from the area around the mounting screw identified as "Bare Metal" in Figure 16. Mount the 3-lug terminal strip with the ground lug under this screw. Doing so ensures that the AC power line ground (the green line cord wire) connects to the chassis.

Remove 1 ½" of insulation from the power cable and strip the last 3/8" of each internal wire. Feed the power cable through the grommet on the rear of the chassis and make connections to the terminal strip following the color code shown in the figure above.

Complete the wiring as shown in the figure above. Use the black and white primary winding of the power transformer T3 for a nominal 120 vac power line voltage. Use the grey and white wires if the AC line voltage is 115 vac. Cap off the unused wire with a wire nut.

The wiring of the SET amplifier is now complete. Before first powering up, it is wise to do some circuit resistive and voltage checks.

Before installing the tubes, do the resistance checks in the table below. Refer to the figure below for the numbered points in the table.

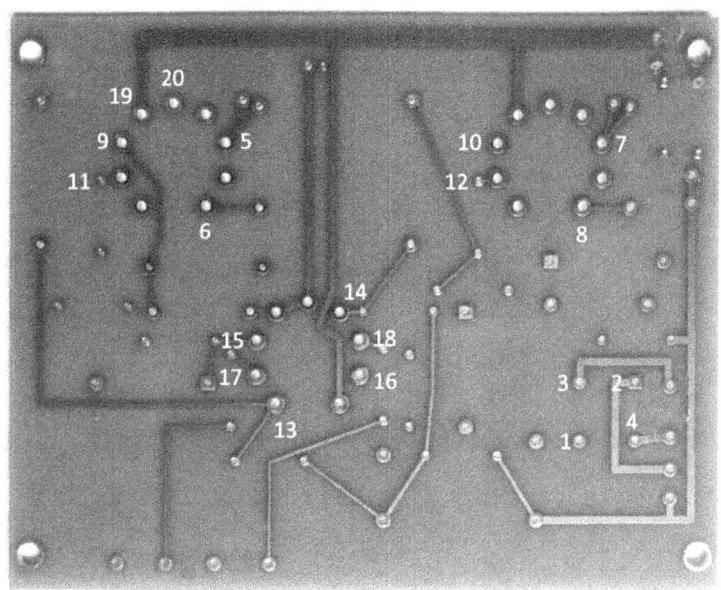

Point-to-Point Resistance* Note

1	G	0.0 Ω	"G" is negative lead of C7 or C8
2	B+	111 Ω	"B+" is positive lead of C7
3	4	172 Ω	
3	G	inf	
B+	G	≥ 100 KΩ	
5	B+	177 Ω	
6	B+	277 Ω	
7	B+	177 Ω	
8	B+	277 Ω	
9	G	270 Ω	
10	G	270 Ω	
11	G	471 KΩ	
12	G	471 KΩ	
13	B+	222.7 KΩ	

14	B+	222.7 KΩ	
15	G	2.7 KΩ	
16	G	2.7 KΩ	
17	G	110 KΩ	Volume control fully clockwise
18	G	110 KΩ	Volume control fully clockwise
19	20	0.4 Ω	
19	G	50 Ω	

* Because of component tolerances, allow up to ±10% variation.

To check the power input circuit, attach the Ωmeter to the blade prongs of the power cord. With the power switch SW1 in the "off" position, the resistance should indicate an open circuit. In the "on" position, resistance should be approximately 14-16 Ω. The round pin on the AC plug should connect to the metal chassis. Neither of the blade pins should connect to the metal chassis.

These resistance checks should identify most circuit problems. If nothing is amiss, install the tubes and proceed with an operational test. With speakers and an audio source connected, apply power, and enjoy your SET amplifier!

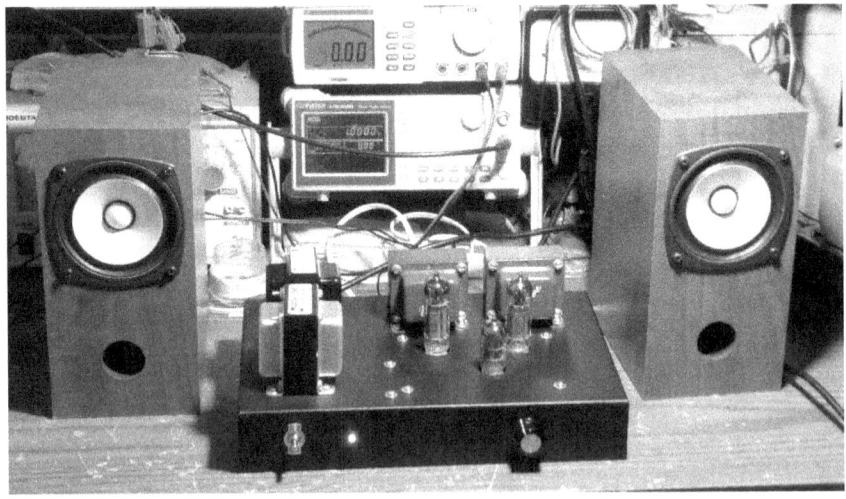

Chapter 15 – Wrapping-Up

In the preceding chapters, we began with the lowly electron and finished with a SET amplifier. We laid the theoretical foundation for SET amplifier design and carried out an example design from start to finish. The knowledge and experience should serve you well as you experiment with other SET amplifier designs.

Perhaps a more powerful SET amplifier would be your goal. You could replace the power output tubes with 6L6-GCs. You could accomplish this with the same size chassis and minimal modification to the design. The voltage amplifier stage would be the same, with only minor component changes in the power output stage (like the self-bias resistor). The power supply would have to deliver more current, possibly necessitating a different power transformer.

Whatever direction you take, I hope you have found this introduction to simple high fidelity with a Single Ended Triode a worthwhile undertaking.

Appendices

Appendix A - Derive Maximum Power Transfer – Linear Case

Consider the power output circuit below.

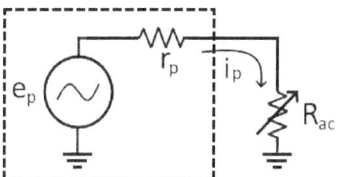

If we assume that the power output tube and circuit are linear, then the following equations apply:

$$P = \frac{(R_{ac}i^2)}{8} \quad i = \frac{e_p}{r_p + R_{ac}}$$

Substituting for i in the first equation, we have

$$P = \frac{e_p^2}{8} R_{ac}(r_p + R_{ac})^{-2}$$

To find the value of R_{ac} that maximizes P, we take the derivative of P with respect to R_{ac}, set it equal to zero, and solve it for R_{ac}.

$$\frac{dP}{dR_{ac}} = \frac{e_p^2}{8}\left[R_{ac} \cdot -2(r_p + R_{ac})^{-3} + 1 \cdot (r_p + R_{ac})^{-2}\right] = 0$$

Multiply both sides by $(r_p + R_{ac})^3$, we have

-2R_{ac} + (r_p + R_{ac}) = 0

R_{ac} = r_p

Power maximum occurs when R_{ac} = r_p.

Appendix B – Derive Max Power Transfer – Non-Linear Case

"Real" power output tubes are non-linear. For this reason, we must represent them graphically, as shown below.

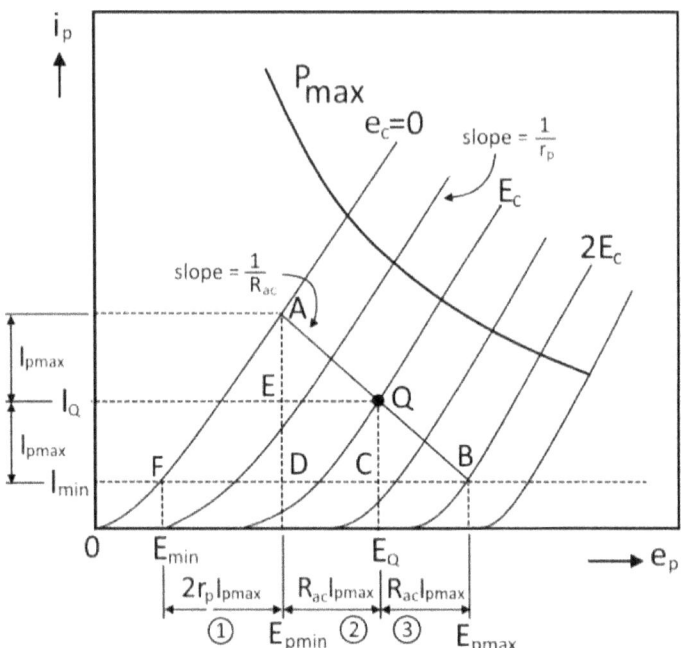

The transfer characteristics from input e_c to either e_p or i_p are not linear, as indicated by the curved plate curves. To mediate this somewhat, we place limits on the load line as follows: (1) $e_c \leq 0$ vdc – point A, (2) $i_p \geq i_{min}$ – point B, and (3) DC operating point Q is exactly halfway between A and B – point C.

To get length 1, we use the triangle ADF shown below.

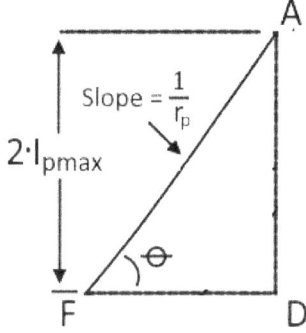

The figure is a right triangle with side AD equal to 2·I$_{pmax}$. The slope of AF equals the tangent of angle theta Θ, which is the ratio of AD to DF. We then can write.

$$\frac{AD}{DF} = \frac{1}{r_p}$$

But AD is 2·I$_{pmax}$, so we have

$$\frac{2 \cdot I_{pmax}}{DF} = \frac{1}{r_p}$$

Solving for DF,

$$DF = 2 \cdot r_p \cdot I_{pmax}$$

DF is length 1.

Moving on to lengths 2 and 3, we use this triangle ABD.

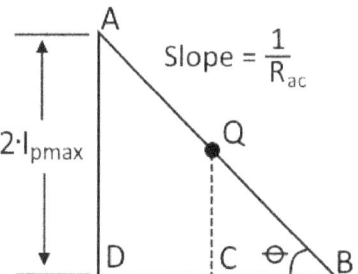

The figure is a right triangle with side AD equal to 2·I$_{pmax}$. The slope of AB equals the tangent of angle theta Θ, which is the ratio of AD to DB. We then can write.

$$\frac{AD}{DB} = \frac{1}{R_{ac}}$$

But AD is 2·I$_{pmax}$, so we have

$$\frac{2 \cdot I_{pmax}}{DB} = \frac{1}{R_{ac}}$$

Solving for DF,

$$DB = 2 \cdot R_{ac} \cdot I_{pmax}$$

Given that DC = CB and DC + CB = DB, we have DC = CB = R$_{ac}$· I$_{pmax}$, which equals lengths 2 and 3.

Then, we can write

$$E_Q - E_{min} = 2 \cdot r_p \cdot I_{pmax} + R_{ac} \cdot I_{pmax} = I_{pmax} \cdot (2 \cdot r_p + R_{ac})$$

Solving for I$_{pmax}$, we have

$$I_{pmax} = \frac{E_Q - E_{min}}{(2 \cdot r_p + R_{ac})}$$

The power output is

$$P = \frac{I_{pmax}^2}{2} \cdot R_{ac} = \frac{(E_Q - E_{min})^2}{2} R_{ac}(2r_p + R_{ac})^{-2}$$

To find the value of R$_{ac}$ that maximizes P, we take the derivative of P with respect to R$_{ac}$, set it equal to zero, and solve it for R$_{ac}$.

$$\frac{dP}{dR_{ac}} = \frac{(E_Q - E_{min})^2}{2} \left[R_{ac} \cdot -2(2r_p + R_{ac})^{-3} + 1 \cdot (2r_p + R_{ac})^{-2} \right] = 0$$

Multiply both sides by (2r$_p$ + R$_{ac}$)$^{-3}$, we have

$-2R_{ac} + (2r_p + R_{ac}) = 0$

$R_{ac} = 2r_p$

Maximum power with the prescribed limits occurs when $R_{ac} = 2 \cdot r_p$.

Appendix C – Non-Linear Distortion

If we introduce a waveform into an amplifier, we want the output to resemble the input in shape exactly. To achieve this exact reproduction, the amplifier's gain must be constant at all input/output levels. When it is not, the result is *non-linear distortion*.

Harmonic Distortion

Non-linear distortion introduces harmonics of the original signal that occur at multiples of the signal frequency (called the fundamental frequency). We often refer to harmonics as either even-ordered or odd-ordered; that is, second, fourth, and so forth harmonics are *even-ordered* and third, fifth, and so forth are *odd-ordered*.

We calculate *the Percent Total Harmonic Distortion* using the formula.

$$THD = \frac{\sqrt{V_2^2 + V_4^2 + \cdots + V_3^2 + V_3^2 + \cdots}}{V_1} \cdot 100\%$$

where V_1 is the RMS voltage of the signal's fundamental frequency, and V_n is the RMS voltage of the nth harmonic.

Triode power tubes are not capable of constant gain across their operating input/output range. We see this as the different spacing of the plate curves across the span of the AC load line. Typically, spacing becomes compressed as the plate voltage swings toward its most positive value. The resulting output waveform has a slight flattening, as shown below. In terms of harmonic distortion, even order harmonics prevail, particularly the second harmonic.

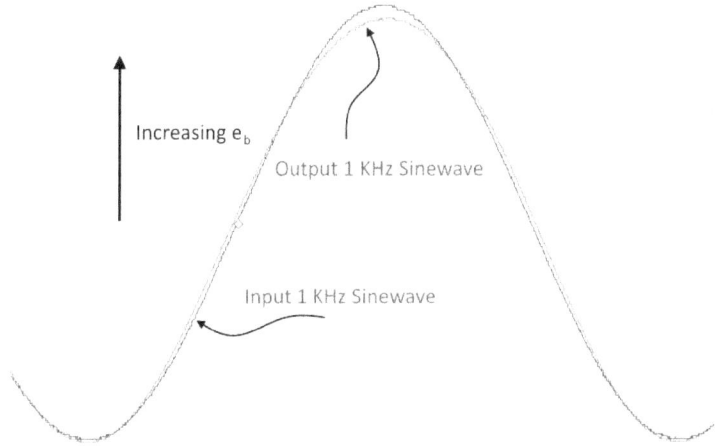

The Set amplifier tested produced 1-watt audio with 3.7% THD (Total Harmonic Distortion) from 1 KHz input signal. 3.5% was due to the second-harmonic distortion with only 0.2%, the third-harmonic and greater. This result is very characteristic of single-ended triode amplifiers.

Visually comparing the two waveforms, we see that 3.7% second-order distortion produces a minor change in waveform shape. It is easy to understand why this level of distortion in a SET amplifier is barely perceptible to the average human listener!

On the other hand, odd-order distortion sharpens the waveform, creating edges and corners that the human ear readily detects and finds highly objectionable. For instance, if we drive our amplifier to the point of clipping the signal, odd-order harmonics suddenly appear, and the waveform takes on a "squared-off" appearance. Eventually, if we drive it hard enough, the signal begins to resemble a square wave.

Even the most casual listener can detect the clipping point with its introduction of odd-order harmonics!

Intermodulation Distortion

A second type of distortion caused by non-linearity *is intermodulation distortion,* or IM distortion, for short. Two or more signals of different frequencies passing through a non-linear amplifier create spurious components at the sum and difference of the frequencies.

For example, suppose we input 1000 Hz and 1500 Hz signals to a SET amplifier; the output consists of these two frequencies plus additional signals at 500 Hz and 2500 Hz. The latter two signals are much smaller but, taken together, can create an IM distortion greater than harmonic distortion, as much as twice or more in percentage.

If the lower frequency is much larger than the upper frequency, the former can modulate the latter, producing readily perceivable distortion. Still, another effect occurs when the low-frequency signal is sub-audible while the difference frequency is not. The result for the listener is a bass sound that is not in the original signal, a sort of "false bass."

IM distortion sounds worse than it is, at least in the author's experience. IM distortion in a SET amplifier is far less noticeable than third-order harmonic distortion in a single-ended pentode amplifier. In fact, the amplifier's IM distortion is frequently covered up by IM distortion caused by the speaker and even human hearing. As far as SET amplifiers are concerned, IM distortion is rarely an issue to contend with.

Appendix D – Calculate Second-Order Distortion

One advantage of triode power tubes is that harmonic distortion is primarily second order. If we assume that it is, we can develop a rather straightforward way to estimate distortion at a given operating point and load line.

Consider the plate characteristic with DC operating point Q and AC load line.

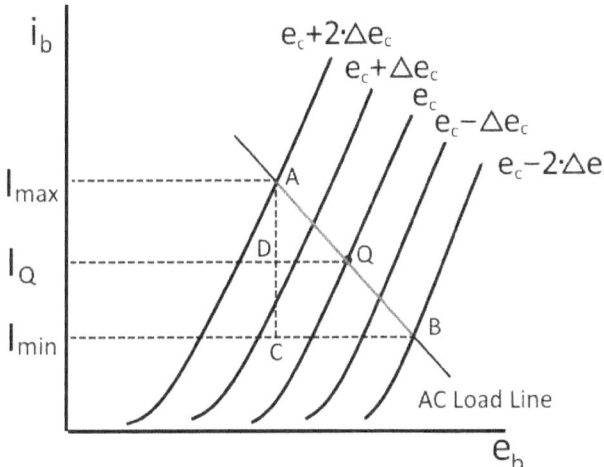

Δe_c is the step change in grid control voltage. We assume the driving voltage $e_i = e_c + 2 \cdot \Delta e_c \cdot \cos(\omega t)$. The control grid voltage swings between $e_c + 2 \cdot \Delta e_c$ and $e_c + 2 \cdot \Delta e_c$, with the output current swinging between i_{max} and i_{min}, respectively.

As this is a triode characteristic, we assume the distorted output current i_p consists of a DC average value plus first and second-order cosine terms.

$$i_p = I_A + \sqrt{2} \cdot [I_1 \cos(\omega t) + I_2 \cos(2\omega t)]$$

I_A is the average DC value of the distorted sinusoid. If distortion is small, I_A is approximately equal to I_Q. I_1 and I_2 are, respectively, the magnitudes of the fundamental and second-order components of the output current. The value ω is the angular frequency of the driving sinusoid signal. Finally, the variable t is time.

We can now develop three equations in three unknowns by substituting values of i_p at three values of ωt, namely 0, $\pi/2$, and π.

At $\omega t = 0$, $\cos(\omega t) = 1$ and $\cos(2\omega t) = 1$. Then $e_i = e_c + 2 \cdot \Delta e_c$. Looking at the figure above, we see that for this e_c, i_p is i_{max}. So, we can write the following equation.

$$I_{max} = I_A + \sqrt{2} \cdot I_1 + \sqrt{2} \cdot I_2 \quad \text{Eqn. 1}$$

At $\omega t = \pi/2$, $\cos(\omega t) = 0$ and $\cos(2\omega t) = -1$. Then $e_i = e_c$. Looking at the figure above, we see that for this e_c, i_p is i_Q. So, we can write the following equation.

$$I_Q = I_A - \sqrt{2} \cdot I_2 \quad \text{Eqn. 2}$$

At $\omega t = \pi$, $\cos(\omega t) = -1$ and $\cos(2\omega t) = 1$. Then $e_i = e_c - 2 \cdot \Delta e_c$. Looking at the figure above, we see that for this e_c, i_p is i_{min}. So, we can write the following equation.

$$I_{min} = I_A - \sqrt{2} \cdot I_1 + \sqrt{2} \cdot I_2 \quad \text{Eqn. 3}$$

Solving equations 1, 2, and 3 simultaneously, we have

$$I_A = \frac{I_{max} + I_{min}}{4} + \frac{I_Q}{2} \quad \text{Eqn. 4}$$

$$\sqrt{2} \cdot I_1 = \frac{I_{max} - I_{min}}{2} \quad \text{Eqn. 5}$$

$$\sqrt{2} \cdot I_2 = \frac{I_{max} + I_{min}}{4} - \frac{I_Q}{2} \quad \text{Eqn. 6}$$

If I_A differs substantially from I_Q, move the AC load line to (I_A, E_Q) and repeat the calculation. Continue this repeat procedure until the new and old values of I_A are reasonably close together. At that point, the approximate percent second-order distortion is the ratio of I_2 to I_1 times 100% or from equations 5 and 6

$$\text{Triode Harmonic Ditortion} = \frac{I_2}{I_1} = \frac{\frac{1}{2}(I_{max} + I_{min}) - I_Q}{I_{max} - I_{min}} \cdot 100\% \quad \text{Eqn. 7}$$

We can carry this one step further by relating this equation to the lengths of line segments AQ and QB in the figure above. Note the following relationships:

$$I_{max} - I_{min} = AC, \text{ the length of one side of triangle ACB} \quad \text{Eqn. 7}$$

$$\frac{1}{2}(I_{max} - I_{min}) - I_Q = \frac{1}{2}[(I_{max} - I_Q) - (I_Q - I_{min})] \quad \text{Eqn. 8}$$

$(I_{max} - I_{min}) = AD$ and $(I_{max} - I_Q) = DC$ Eqns. 9 and 10

Substituting Eqns. 9 and 10 into Eqn. 8, we have

$$\frac{1}{2}(I_{max} - I_{min}) - I_Q = \frac{1}{2}[AD - DC] \quad \text{Eqn. 11}$$

From trigonometry, we can show that

$$\frac{AD-DC}{AC} = \frac{AQ-QB}{AB} \text{ and}$$

$$AD - DC = \frac{AQ-QB}{AB} \cdot AC = \frac{AQ-QB}{AQ+QB} \cdot AC \quad \text{Eqn. 12}$$

Making the appropriate substitutions of Eqns. 7, 11, and 12, we have

$$Triode\ Harmonic\ Distortion \sim \frac{AQ - QB}{2(AQ + QB)} \cdot 100\%$$

With rearranging, we have

$$Triode\ Harmonic\ Ditortion \sim \frac{\frac{AQ}{QB} - 1}{2(\frac{AQ}{QB} + 1)} \cdot 100\%$$

If we plot THD by the ratio AQ/BQ, we can conveniently estimate harmonic distortion. Measure AQ and BQ for a given AC load line, calculate their ratio, and then read the estimated harmonic distortion.

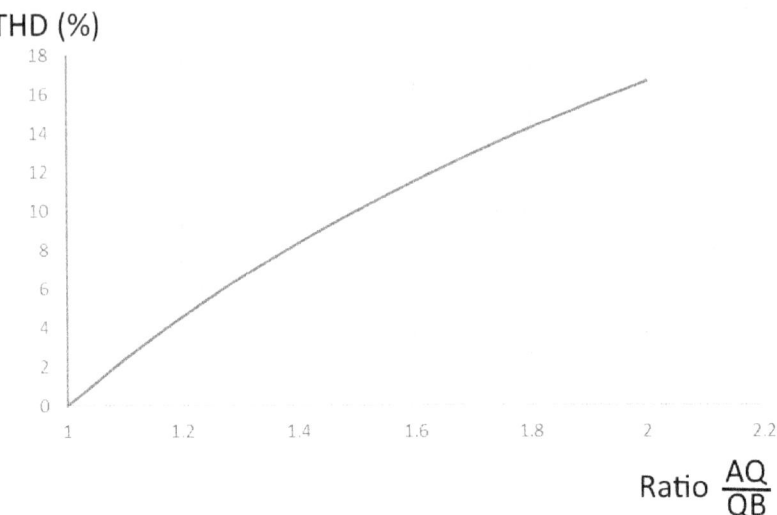

Appendix E – Frequency Distortion

Most sounds consist of a mixture of frequencies, not just a single tone. To achieve undistorted output, the SET amplifier must pass all the sound's frequency components without increasing or decreasing any of their amplitudes. When an amplifier's gain varies with frequency, the result is that a sound's mixture of frequencies amplifies differently, reducing the fidelity of reproduction. We call this *frequency distortio*n.

SET amplifiers have limited frequency response, generally meaning that there is a mib-band of frequencies that exhibit a constant mid-band gain, with that gain falling off both below and above the mid-band. If all the components of a sound fall within the mid-band, frequency distortion is minimal. When this is not the case, the output consists of some frequency components whose attenuated amplitudes result in a reproduced sound that lacks the fidelity of the original.

For SET amplifiers, we define the mid-band defined as the frequency range over which the gain differs by less than some amount, often 1 db. We use such a range as the amplifier's frequency response specification. A minimum for "good fidelity" is 30 to 15,000 Hz. For "high fidelity," we generally expect 20 to 20,000 Hz.

At lower frequencies, the amplifier amplifies sufficient harmonics of the fundamental to produce the richness of a sound reasonably well. But, at higher frequencies, the harmonics are attenuated as the amplifier's gain falls off at the upper end of its frequency response range. The figure below shows this effect in that the lack of high-frequency response rounds the sharp corners of a 5 KHz square wave in the amplifier output.

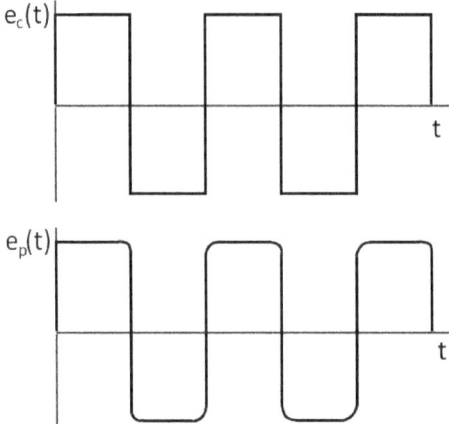

The attenuation of harmonics at 15 KHz and above prevents the reproduction of the sharp corners of the square wave.

In practice, the greatest factor affecting frequency distortion is the audio output transformer. Its design and quality of materials affect the frequency response of the amplifier both above and below mid-band.

Connected with frequency distortion is *phase distortion*. If the amplifier's parameters cause differing phase relationships between signal components, this result is a misshaping of the output signal. Phase distortion occurs most often below and above mid-range. Listening tests appear to indicate that phase distortion is rarely perceivable in complex sounds like music. As with frequency distortion, the best remedy is to aim for the widest mid-range frequency response possible!

Appendix F – Transient Response

At low frequencies below 100 Hz, the SET amplifier may have difficulty reproducing sharp sounds. Several factors play a role in this. The audio output transformer may have limited response in this low range of frequencies. The power supply may not have enough reserve to maintain the plate supply voltage as the frequencies dip below the replenishing cycle dictated by the power line frequency and type of rectifier. Lastly, the capacitive coupling between the voltage amplifier and the power output stage fails at low frequencies. Combining all three, we end up with a transient response like that shown below.

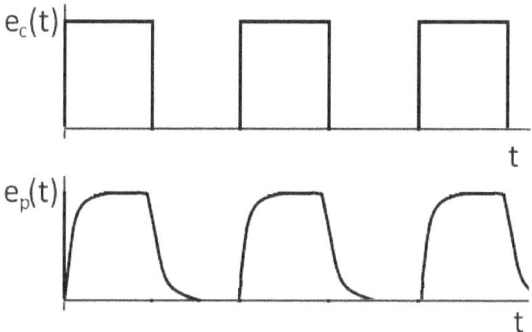

The sharp rise times of the input square wave are compromised to make it look more like a capacitor charging and discharging. At frequencies below 20 Hz, the output waveform can become so distorted that it is hardly recognizable.

The redeeming feature is that little sound reproduction relies on such low-frequency content. Here again, factors like the speaker and human hearing often limit reproduction in these low-frequency ranges far more than the SET amplifier.

Appendix G – The Miller Effect

The basic mechanism that produces the *Miller Effect* is relatively simple to analyze. Consider the amplifier below with a gain of -A and an impedance connected from the output to the input.

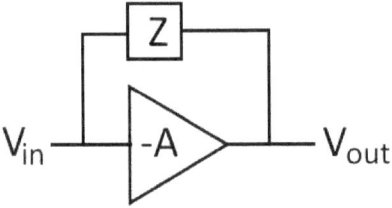

Assume the input voltage V_{in} changes by ΔV. The change in voltage across the impedance is

$\Delta V_z = \Delta V - (-A*\Delta V) = \Delta V(1+A)$

Thus, the change in current in the feedback impedance is

$\Delta I = \Delta V_z/Z = \Delta V(1+A)/Z$

The effective impedance seen by the input is

$Z_{eff} = \Delta V/\Delta I = Z/(1+A)$

If the feedback impedance Z is a capacitor, then it looks like a capacitor that is (1+A) times as large.

Appendix H – Resistance vs. Impedance

When we measure the effect of a load on a DC voltage and current, we use the term *resistance*. When we measure the effect of a load on an AC voltage and current, we use the term *impedance*. The unit of measure for both is Ω. An effective way to remember the difference is to make a mental note that circuits "resist DC" and "impede AC."

Further, a load that responds equally to DC and AC is said to be *resistive*. A resistor is obviously a resistive device. A load that responds differently to DC and AC is said to be *reactive*. An inductor is a reactive device. Reactive loads can be frequency dependent; that is, the impedance of a reactive load may vary with frequency. For an inductor, the magnitude of reactance is 1/(2πfL), where f is the sinusoid frequency, and L is the inductance.

Though technically, resistance and impedance are not the same, their behaviors are similar and follow many of the principles. Consider the example circuit below, which consists of an equivalent tube circuit driving a reactive load.

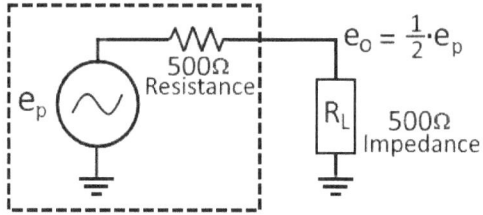

For a specific AC frequency, the plate resistance is 500 Ω, and the load impedance is 500 Ω. The voltage divider principle applies to this mix of DC resistance and AC impedance, making e_o exactly half of e_p; that is, if e_p is 10 vpp, then e_o is 5 vpp.

We could use this technique to find the value of an unknown coil's inductance. With the circuit above, we would change the frequency of e_p until e_o equals one-half e_p. Suppose this frequency is 100 Hz; then we calculate L as shown below.

$$L = \frac{1}{2 \cdot \pi \cdot 100 \cdot 500} = 3.16 \; \mu H.$$

In amplifier design, we encounter impedance when matching the speaker impedance to a specified load resistance value for the power output tube. Here again, we can mix resistance and impedance so long as we are considering audio frequencies.

We should also note that devices, in general, can be themselves mixtures of resistive and reactance behaviors, which is where complex numbers come into play. We are now way beyond what we need for our SET amplifier work. The internet has many good introductions to AC circuit analysis, which is how reactance and complex numbers come together. See, for example, http://www.physicsbootcamp.org/Circuit-Analysis-Using-Complex-Numbers.html.

Appendix I – Voltage Amplifier Design – General

To design a triode voltage, we first select the tube, knowing the desired voltage gain Av = 60. The 12AX7 readily provides voltage gains from 25 to 80 with low noise and harmonic distortion, so we selected it for our example. The figure below shows the basic voltage amplifier circuit.

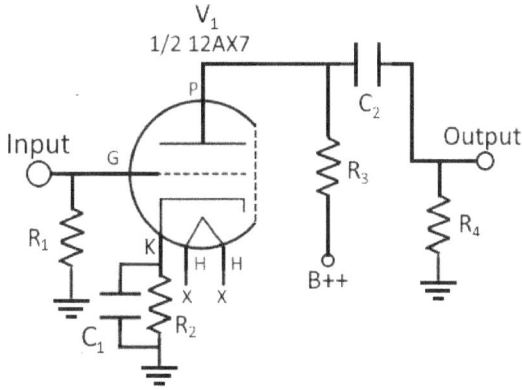

The recommended value for R_1 is 100K Ω, although the input specification could call for a specific load resistance, either higher or lower. If R_1 is the volume control, then use an audio taper 100K Ω potentiometer with the wiper connected to the control grid G.

Next, assume a plate supply voltage of 250 vdc. To give the output audio signal maximum symmetrical swing, set the quiescent plate voltage E_Q at half of the B++, so E_Q = 125 vdc. The 12AX7 specs show a suggested V_c = -1.0 vdc for 100 vdc plate voltage. The same would be a reasonable value for our slightly higher E_Q of 125 vdc. Plot the quiescent point Q on the 12AX7 plate characteristics at the intersection of plate curve V_c = -1 vdc and E_p = 125 vdc. See the figure below.

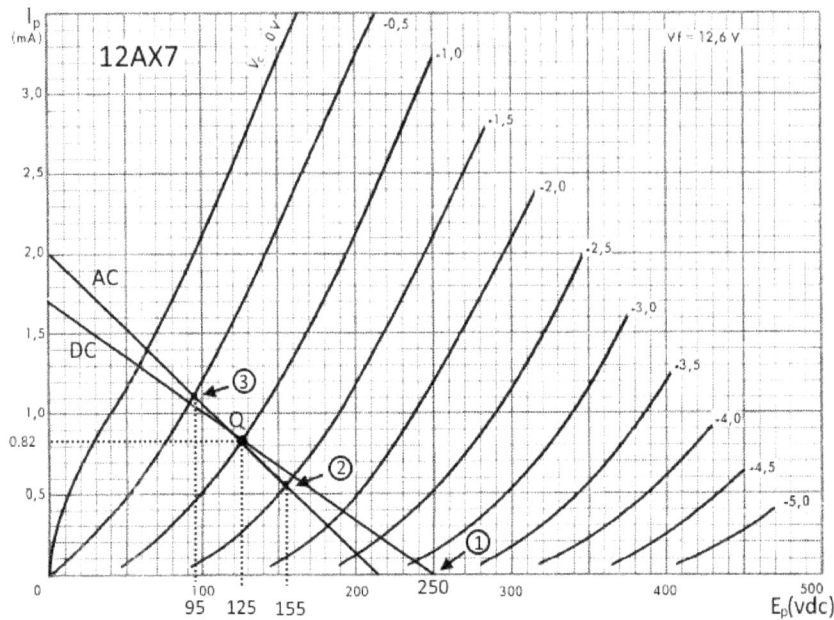

Next, draw a line from B++ = 250 vdc (point 1) through point Q to the plate current axis, where it intersects at 1.7 ma. We call this line the DC load line. If the DC plate load resistance is R, then 250 / R = 1.7 mA and R ≈ 150K Ω. Because the self-bias resistor R_2 is much smaller than plate resistor R_3, we ignore it and make R_3 = R = 150K Ω.

Voltage gain at this DC operating point is ~75. However, at audio signal frequencies, capacitor C_2 is a short circuit, and resistor R3 is parallel with resistor R_4. The effect of this is to increase the slant of the load line, pivoting it about point Q and decreasing voltage gain. For design, we want to decrease to 60.

Our approach is to start with the desired voltage gain and work backward to calculate R_4. In the figure above, mark points 2 and 3 where the plate voltage 125 ± 30 vdc intersects control grid steps -1 ± 0.5 vdc. This range establishes an output voltage of 60 vpp for an input voltage of 1 vpp, which gives us the required $A_v = 60$. Now, draw the line through points 2, 3, and Q. This is the AC load line.

The AC load line intersects the plate current axis at 2.0 ma. The AC load resistance is, therefore, $125/(2.0-0.82)$ KΩ = 106K Ω. That is, the parallel combination of R_3 and R_4 should be 106K Ω for $A_v = 60$. Using the parallel resistance formula, we have

$$106 = \frac{150 \cdot R_4}{150 + R_4}$$

$$106 \cdot 150 + 106 \cdot R_4 = 150 \cdot R_4$$

$$44 \cdot R_4 = 15.900$$

Solving for R_4, we get 361K Ω. Using the standard value of 390K Ω for R_4 raises the voltage gain slightly.

Self-bias resistor R_2 is simply $-1 / 0.82$ K$\Omega \approx 1200$ Ω. As previously determined, for a frequency response of 20 to 20,000 Hz, make $C_1 = 47$ µF and $C_2 = 0.1$ µF. The figure below shows the completed voltage amplifier design.

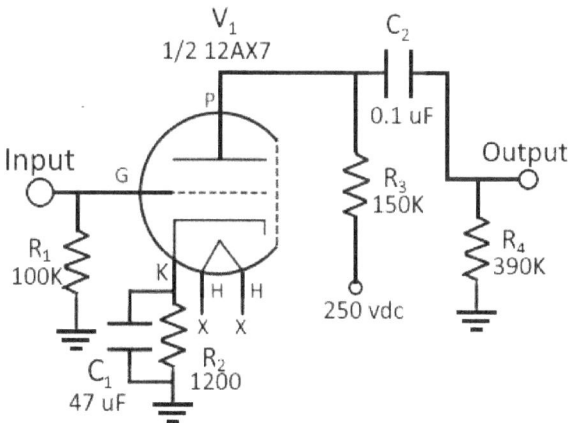

Appendix J – SET Amplifier Parts List

AES – Antique Radio Supply https://www.tubesandmore.com/
EDCORE – EDCORE USA https://edcorusa.com/
Jameco – Jameco Electronics https://www.jameco.com/

B1 Bridge Rect   AES P-QBR-34

C1 50uF 50 vdc   AES C-SA50-50

C2 0.1uf 630 vdc   AES C-TD1-630

C3 250uF 25 vdc   AES C-ET250-25-MOD

C4 50uF 50 vdc   AES C-SA50-50

C5 0.1uF 630 vdc   AES C-TD1-630

C6 250uF 25 vdc   AES C-ET250-25-MOD

C7 100uF 450 vdc   AES C-ET100-450

C8 47uF 450 vdc   AES C-ET47-450

C9 0.047uF 400 vdc   AES C-PD047-400

F1 1/2 A 250 vdc   AES F-Z3AG-S0D5

J1 RCA jack   AES S-H267R

J2 Dual Binding Post Jameco GBRR2-R

J3 RCA jack   AES W-SC-3501FR

J4 Dual Binding Post Jameco GBRR2-R

L1 5 H   AES P-C194F

P1 AC power cord   AES S-W132

R1 100K audio taper   AES R-V38-2X100KA

Component	Value	Part
R2	2.7K 1/2 w	AES R-A2D7K
R3	220K 1/2 w	AES R-A220K
R4	470K 1/2 w	AES R-A470K
R5	1K 1/2 w	AES R-A2D1K
R6	270 1 w	AES R-B270
R7	100 1 w	AES R-B100AES R-B100
R8	100k audio taper	AES R-V38-2X100KA
R9	2.7K 1/2 w	AES R-A2D7K
R10	220K 1.2 w	AES R-A220K
R11	470K 1/2 w	AES R-A470K
R12	1K 1/2 w	AES R-A2D1K
R13	270 1 w	AES R-B270
R14	100 1 w	AES R-B100
R15	2700 1/2 w	AES R-A2D7K
R16	100K 1/2 w	AES R-B100K
R17	100 1/2 w	AES R-A100
R18	100 1/2 w	AES R-A100
R19	1K 1/2 w	AES R-A2D1K
R20	1K 1/2 w	AES R-A2D1K
SW1	SPST switch	AES P-H495-T
T1	5K Pri - 8 Sec	EDCOR GXSE10-5K
T2	5K Pri - 8 Sec	EDCOR GXSE10-5K

T3 Pwr XFRM   AES P-T269AX

V1 12AX7   AES T-12AX7-S-JJ

V2 6BQ5   AES T-EL84-JJ

V3 6BQ5   AES T-EL84-JJ

Misc Parts

4x 1/2" 6-32 standoffs   AES S-H170

Fuse holder   AES S-H259

Grommets 3/8" (x7)   AES P-GROM-38

Grommets 1/4" (1)   AES P-GROM-14

6-32 x 3/8" screws   AES S-HS632-38

6-32 washers   AES S-HW6

6-32 lock washers   AES S-HLW6

6-32 nuts   AES S-HHN632

5-lug terminal strip x 2   AES P-0501H01

3-lug terminal strip   AES P-0301H

Fuse holder   AES S-H259

4-40 x 1/4" screw   AES S-HS440-14

4-40 nut   AES S-HHN440

Shielded cable   Jameco 228401*

* Make from audio patch cable

Appendix K – ExpressPCB SET Amplifier Schematic

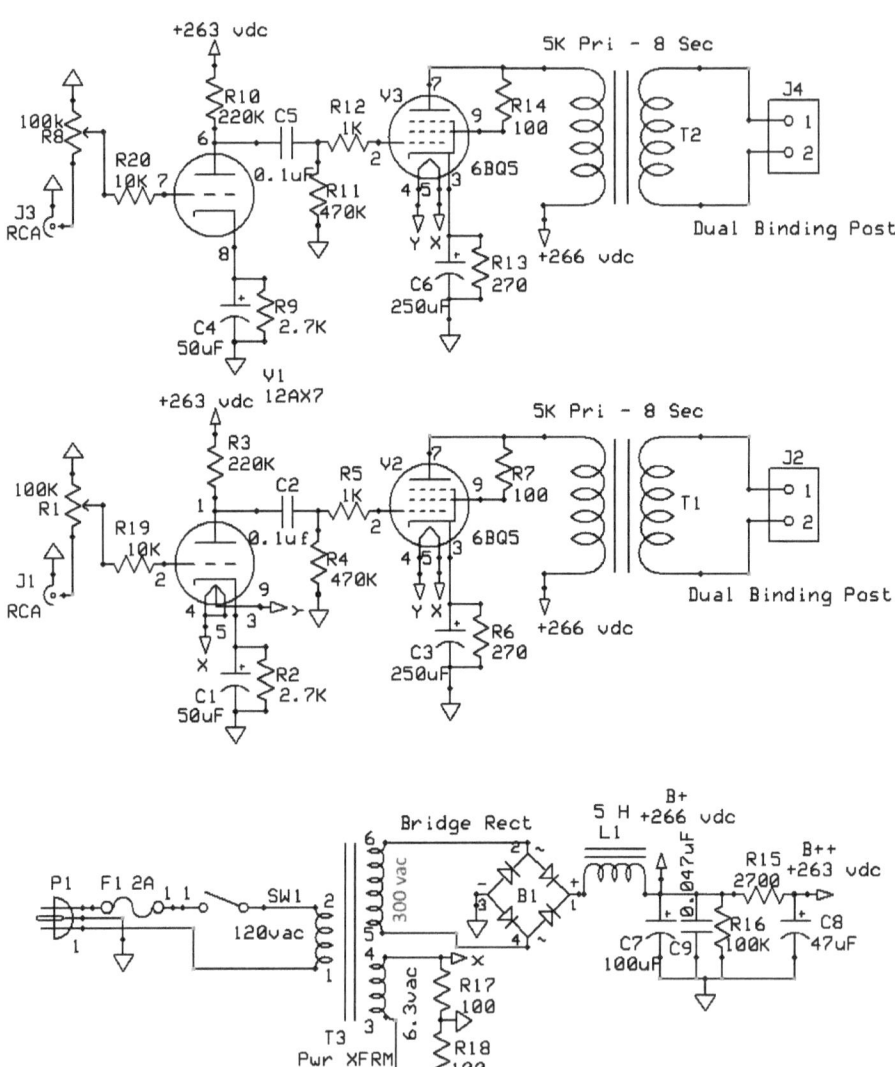

www.ingramcontent.com/pod-product-compliance
Lightning Source LLC
Chambersburg PA
CBHW052354220526
45465CB00003BA/1103